# CONTEMPORARY LITERARY LANDSCAPES

# Contemporary Literary Landscapes

## The Poetics of Experience

DANIEL WESTON
*University of Greenwich, UK*

Routledge
Taylor & Francis Group
LONDON AND NEW YORK

First published 2016 by Ashgate Publishing

2 Park Square, Milton Park, Abingdon, Oxfordshire OX14 4RN
52 Vanderbilt Avenue, New York, NY 10017

*Routledge is an imprint of the Taylor & Francis Group, an informa business*

First issued in paperback 2018



**British Library Cataloguing in Publication Data**
A catalogue record for this book is available from the British Library

**The Library of Congress has cataloged the printed edition as follows:**
Names: Weston, Daniel.
Title: Contemporary literary landscapes: the poetics of experience / by Daniel Weston.
Description: Farnham, Surrey, UK; Burlington, VT: Ashgate, [2016] |
Includes index.
Identifiers: LCCN 2015033077 | ISBN 9781472474650 (hardcover) |
Subjects: LCSH: English literature—20th century—History and criticism—Theory, etc. | Place (Philosophy) in literature. | Setting (Literature)
Classification: LCC PR471.W47 2016 | DDC 820.9/0091—dc23
LC record available at http://lccn.loc.gov/2015033077

ISBN: 978-1-4724-7465-0 (hbk)
ISBN: 978-0-367-14053-3 (pbk)

# Contents

# Acknowledgements

I wish to thank colleagues and friends who have provided invaluable support, advice, and guidance throughout this project. I am particularly grateful to those who helped me right from the genesis of this work in doctoral study at the University of Nottingham: Julie Sanders, David James, and Neal Alexander. Thanks also to others who helped me from this early period – Dominic Head, Andrew Thacker, Sean Matthews, David Cooper, Tim Cresswell – and those who have assisted me more recently in bringing the project to fruition – David Kennedy, Janet Clare. I wish to thank my friends Louise Chamberlain, Sam Haddow, Dan Hunt, Dave Peplow, and Paul Whickman, my colleagues and former colleagues in the English departments at the University of Hull and the University of Nottingham, Ann Donahue and others who have worked on this book at Ashgate, as well as the anonymous reader of the manuscript.

My parents, Valerie Faulks and Patrick Weston, as well as John Faulks, Hannah Woolley, Ben Weston, David Woolley, and all of my family have always supported me and my work in academia. I am thankful for them all. Richa Dwor helped me in very many ways. Peggy Naumann helped me get this finished. Thank you.

Parts of this book have appeared elsewhere. A version of Chapter 1 was published as 'The spatial supplement: landscape and perspective in W.G. Sebald's *The Rings of Saturn*', *Cultural Geographies* 18.2 (2011): 171–86. A version of Chapter 2 was published as '"I know this labyrinth so well": Narrative Mappings in the Poetry of Ciaran Carson', *Poetry and Geography: Space and Place in Post-War Poetry*, ed. Neal Alexander and David Cooper (Liverpool: Liverpool University Press, 2013), 105–19. I thank Sage Publications and Liverpool University Press for allowing me to reproduce work here in modified form.

# Introduction
# Articulating Place

Writing landscapes inevitably occurs in dialogue with a long textual and pictorial tradition, but first-hand experience also provides key stimuli to many writers' accounts. In response, scholarship has played out a tension between genre and experience. Some cultural geographers are concerned that self-perpetuating literary tropes marginalize practical engagements. They have elaborated other, non-representational models for understanding landscape. This monograph employs a comparative lens to intervene in these debates by arguing that representation and experience are not competing paradigms for landscape. Rather, in the hands of contemporary writers they are complementary forces building composite articulations of place. Five case studies match a writer to a mode of apprehending place – W.G. Sebald with picturing, Ciaran Carson with mapping, Iain Sinclair with walking, Robert Macfarlane with engaging, Kathleen Jamie with noticing – to draw out a range of sites at which representation and experience interact. Key points are twofold: firstly, interaction between traditions of landscape writing and direct experience of landscapes is reciprocal, both influencing the other; secondly, writers increasingly deploy style, form, and descriptive aesthetics to recover the experience of place in the poetics of the text itself. Authors today are often not writing place *per se*, but writing the experience of place.

This monograph is firmly grounded in the interdisciplinary enterprise that has characterized much recent literary analysis. It looks to critical paradigms from cultural geography to devise innovative new ways of thinking about landscape writing and to assess the form and politics of place, as well as the poetics and narratives of spatial texts. It responds to those forms of cultural geography that seem least compatible with literary studies: phenomenological versions of landscape that stress embodied practice over textual traditions. This work contributes to critical debates via close reading. By conferring central importance on sustained textual analysis, hybrid methodological innovations emerge, whilst the monograph also samples contemporary landscape writing across fiction, poetry, literary non-fiction, and texts that blur these distinctions. It contends that emergent landscape writing shuttles across generic boundaries, reflecting the fact that the landscapes traversed are built out of a combination of real and imaginary sources. The book therefore considers the important issue of genre in the study of contemporary literature. Its particular field of interest offers a paradigm for important changes informing the contemporary literary scene at large. Further, it responds to perennial declarations of the death of some genres and the birth of others by noting the hybridity that defines much contemporary writing.

This introductory chapter forges the critical position from which subsequent chapters' extended readings are launched. It traces dialogues between the different players in the development of landscape studies, focusing on literary criticism and cultural geography but also considering interventions from environmental history, anthropology, art history, performance studies, and critical theory. The chapter maps the trajectory of interest in landscape and place from a shared field of study with the retention of discrete methodologies, towards the current situation where methodologies mutually inform one another. However, this chapter also makes the case that a truly comparative and combinatory approach has not yet been fully achieved. It details the significant impact of non-representational theory – deriving from the philosophies of Wittgenstein, Heidegger, and Merleau-Ponty – on cultural geography and the turn to embodied practice that has resulted from a reaction to overdetermined readings of places and texts. The two terms 'representation' and 'experience' have often been conceived in an oppositional relationship, with the former fixing universalizing meanings of place that the latter works to overcome in particularized sensory engagement. I build upon recent trends to undermine this distinction, suggesting instead that representation and experience are interwoven in the articulation of places and that the recognition of their complex interaction can inform a study of literary landscapes. Literary criticism and non-representational theory are not incommensurate, as is often thought. The dynamics and poetics of literary form do not arrest embodied experience but work towards its recuperation. This methodological intervention is explained through a short study of the idea of 'articulating' place as it appears in landscape poems by John Burnside and Thomas A. Clark. In this reading practice, the introduction thus proceeds to do in miniature what subsequent chapters do at length.

## Interdisciplinary Dialogues

The establishment of what Peter Brooker and Andrew Thacker call 'a fittingly transdisciplinary vocabulary' has been one of the most pressing tasks of literary geography over the last decade or so (5). In their editorial to a 2005 special issue of *New Formations* entitled 'The Spatial Imaginary', Richard Phillips and Scott McCracken noted that several of the articles included were 'originally given as papers at a conference session, "Textual Spaces, Spatial Texts", at the Royal Geographical Society in London in 2004' (7). The journal issue thus signified a key intervention summarizing the opportunities emerging out of interdisciplinary dialogue between literary criticism and cultural geography, but also the challenges faced therein. 'Such dialogues', Phillips and McCracken noted, 'are not, of course, without disagreement. Cross disciplinary misunderstandings are the inevitable consequences of differences in perspective, language and values: of different critical literacies' (8). Miles Ogborn's article from the same issue was more circumspect, suggesting that 'it is not yet clear whether these different versions of what it means to investigate textual geographies or the geographies of texts can actually speak productively to each other' (149). The same concerns have been reiterated recently in Patricia L. Price's recommendation that her fellow cultural geographers must avoid dilettantism – 'dabbling on the surface of something that seems relevant or interesting, without a thorough training in that something' – and 'give the wide range of approaches we adopt their full intellectual due' (204). Equally, the same warning might be addressed to literary critics in relation to the use they make of geographical concepts. Angharad Saunders observes that 'against a background where the limits of mimetic and metaphorical studies have been widely pronounced upon, literary geography has been subject to some profound epistemological transformations' (436–7). Literary critics have not always fully taken on board the responses to these limitations that geographers have developed, preferring instead to cherry-pick some ideas and leave aside those that are more difficult to reconcile to textual criticism.

However, more recently still, Douglas Richardson has written more positively that the 'resurgence of intellectual interplay between geography and the humanities … has been nothing short of remarkable. Ideas, terminology, and concepts … now permeate both academic and popular cultures as conceptual frameworks, methodologies, and core metaphors' (xix). Richardson's celebration of genuine interdisciplinarity ('interplay') emphasizes the 'reciprocal process' that emerges out of scholars 'productively transgressing the bounds of traditional categories' (xix–xx). Indeed, if terminological concerns do persist, the 'cultural turn' in geography and the 'spatial turn' in arts and humanities disciplines have led to fruitful exchanges. The publication in 2011 of *Envisioning Landscapes; Making Worlds: Geography and the Humanities*, edited by Stephen Daniels et al. (to which Richardson's statement above appears as a foreword), and of *Geohumanities: Art, history, text at the edge of place*, edited by Richardson et al. in the same year, are substantial and wide-ranging collections offering testament to this fact. In their introduction to the latter volume, the editors acknowledge contributors' 'significantly different training', but suggest that the 'fortuitous convergence of intellectual traffic outlines a distinctive scholarly terrain and emerging zone of practice' (Richardson et al. 3). Another edited volume, *Process: Landscape and Text*, edited by Catherine Brace and Adeline Johns-Putra (2010), focuses the same dialogues specifically on literary texts. Whilst edited volumes bring together multiple methodologies, my aim is to sustain ideas over a number of case studies.

Amongst a diverse range of possible methodologies and research aims that fall under the broad rubric of literary geography, Andrew Thacker's formulation, again from an article in the *New Formations* special issue, has been that which literary scholars have taken up most often. Thacker describes 'a critical literary geography' as 'the process of reading and interpreting literary texts by reference to geographical concepts' ('Critical Literary Geography' 60). In this instance, the adjective 'critical' is included 'to stress the distance from an effortless mapping of represented landscapes in literary texts, and to raise more complex questions about space and power, and how space and geography affect literary forms and styles' (60). My own work shares this primary focus on the interpretation of literary texts and the commitment to describing and accounting for their formal and stylistic characteristics. To undertake this kind of analysis through the lens of geographical concepts is, as Thacker notes, to exceed any simple 'mapping of represented landscapes', and has implications that extend beyond literary studies and are of interest to cultural geographers. Thacker suggests that his model is oriented chiefly towards 'what literary and cultural critics can learn from a fuller engagement with theorists of space and geography, rather than the reverse' ('Critical Literary Geography' 56). Others, such as Sara Blair, have formulated a different relationship by asking: 'what precisely do literary critics and historians, no less than theorists, have to contribute to new understandings of spatial practices, of the production of spatial and social differences, and of space, time, and nature as material frames for everyday life?' (549). This study is closer to Thacker's than Blair's model of dialogue in its primary interest in interpreting texts.

That said, whilst my study is engaged in part with a body of geographical work similar to that suggested by Thacker, it also aims to deal more centrally with phenomenological and experiential accounts of place currently defining research agendas in cultural geography. Thacker describes the usefulness (for literary critics) of geographical concepts drawn from Martin Heidegger, Gaston Bachelard, Henri Lefebvre, Michel Foucault, Michel de Certeau, and David Harvey (*Moving Through Modernity* 13–45).[1] Of the other monographs in this area, most are similarly attuned to literary criticism's dialogues with spatial theory and cultural geography broadly conceived. Thus, Ian Davidson's *Ideas of Space in Contemporary Poetry* (2007) acknowledges drawing heavily on Lefebvre and on Doreen Massey 'to provide an intellectual background to the book' but is also indebted to most of the other names in Thacker's sketch too (2). Davidson's subsequent work in *Radical Spaces of Poetry* (2010) turns further in this direction, drawing on 'a range of philosophers and cultural theorists' but not (explicitly) cultural geographers (7). Roger Ebbatson's more recent *Landscape and Literature 1830–1914* (2013) 'seek[s] to respond to elements of phenomenological thought and eco-criticism' but is 'primarily motivated by, and indebted to, Critical Theory as adumbrated by the Frankfurt School' (3). In these examples, spatial theory is the popular bedfellow for interdisciplinary literary studies. The difference in the frame of reference for my own study from that of Thacker and others is not to suggest a better alternative, but rather to acknowledge the quality of work already available in this area and to move towards a different kind of literary geography. Where others have been reluctant to approach a seemingly contrary set of non-representational theories, I hope to demonstrate the potential they unlock for the analysis of landscape writing.

[1] Robert T. Tally Jr.'s *Spatiality*, a primer on this area of study for literary scholars, broadly concurs with Thacker's sense of the important theorists, focusing at length on Michel Foucault, David Harvey, Edward Soja, Erich Auerbach, Georg Lukács, and Fredric Jameson. Other work similarly framed, in addition to that subsequently discussed, includes: Eric Bulson, *Novels, Maps, Modernity: The Spatial Imagination 1850–2000*; David James, *Contemporary British Fiction and the Artistry of Space: Style, Landscape, Perception*; Bertrand Westphal, *Geocriticism: Real and Fictional Spaces*.

Charting key developments in geographers' engagements with literary texts is instructive in pointing up where differences have arisen more recently. Early constructions of the task at hand from these quarters retain a geographical methodology and turn it towards a literary object of study (whilst leaving out literary critical modes of analysis). For example, Douglas Pocock's 1981 construction of the interchange is between literature as 'an artistic creation' and human geography as 'an art or social science', and encompasses 'the search for geography *in* literature, treating writings as a literary quarry from which to construct a more general literary topography' (*Humanistic Geography and Literature* 9, 13).[2] In turn, a strongly demarcated distinction between the real landscapes of geographers and their imaginary counterparts dealt with by literary critics facilitated critiques of literary geography, citing its supposed 'metaphorical appropriation' of geographical concepts (Ogborn 145). It is with these interpretations of the interface between two disciplines that more recent work by literary critics *and* cultural geographers has taken issue.

The means by which this impasse is overcome here might be said to emerge from a suggestion, in embryonic form, in Pocock's writing. 'Setting', he finds, 'achieves existential significance in literature and life through events of experiences. It is the latter which converts setting into place, thereby linking environment and person. Place emerges through experience and in turn symbolizes that experience' ('Geography and literature' 90). Whilst this statement outlines a more fruitful area of exchange, it still illustrates the retention of a methodology defined almost wholly within the borders of the discipline of geography. Marc Brosseau, writing in the same journal as Pocock in 1994, finds that 'the legitimacy of such a relationship to literature is based on a conception of literature as the transcription of a concrete experience' ('Geography's literature' 338). Brosseau takes issue with the simplistic sense of literature's role, neglecting its mediating capacity, that is implied in the idea of 'transcription' of 'concrete' (that is, presumably, 'real') experience. Brosseau's understanding is more nuanced, suggesting that 'it is difficult to assume that we can find this experience in its raw state in literature' (340). If geographers use literature only 'as a source to verify geographical hypotheses', in the instrumental sense detailed above, 'the literary text, in its density and difference, … is obliterated'. Conversely:

> If geographers undertake to entertain a relationship that is more dialogical with a literary text, they cannot overlook the specificity of its form (broadly defined) and of its singular use of language in order to be sensitive to the particular way it generates another type of geography, to the particular way it writes people and place, society and space. (347–8)

---

[2] Other early versions of the relationship include Fabio Lando, 'Fact and fiction: Geography and literature; A bibliographic survey'; William E. Mallory and Paul Simpson-Housley, eds, *Geography and Literature: A Meeting of the Disciplines*; Douglas C.D. Pocock, 'Geography and literature'. For a discussion of the issues at stake and problems with work of this period, see John Kerrigan, 'The Country of the Mind'.

In a further article that, as a paradigmatic case study, might be considered something of a companion piece, Brosseau undertakes this kind of analysis of a specific literary text: John Dos Passos's *Manhattan Transfer* (1925). There, he examines 'the possibilities of acknowledging the *agency* of the text inasmuch as it defines models of readability that cannot be neglected in the appreciation of its specific geography'. A commitment to the fact that 'in order to interpret the city *in* the text, one must examine the geography of the text', leads Brosseau to 'draw attention to the more "technical" aspects of the novel as they help to shed some light on discursive form and the city' ('The city in textual form' 90, 93, 95).

For Sheila Hones, Brosseau's approach 'indicated a major shift in emphasis and, by implication at least, a greatly increased interest in the analytic methods of literary criticism'. Whilst he 'maintained the configuration of the interface in terms of the relationship between geography as a discipline and the literary text, he insisted on the value of the close reading techniques characteristic of literary studies in analysing the subversive power of fiction' (1304). The 'major shift' that Hones identifies was arguably already occurring in the development of a cultural geography invested in close reading at this time and since. Indeed, cultural geographers and others were already introducing new ideas of narrative and form that have led to a reassessment of literary critical commonplaces on these topics in a spatial context. For example, it was uncontroversial for William Cronon, writing before Brosseau's article, to state that environmental historians 'maintain a powerful commitment to narrative form', and to urge upon them 'the task of telling not just stories about nature, but stories about stories about nature' ('A Place for Stories' 1349, 1375). Elsewhere he suggests that 'the special task of environmental historians is to tell stories that carry us back and forth across the boundary between people and nature to reveal just how culturally constructed that boundary is – and how dependent upon natural systems it remains' ('Kennecot Journey' 33). Literary scholars have profited from Cronon's observation that a narrative of place signifies 'neither a historical accident nor a case of geographical determinism', but rather emerges out of a combination of human and non-human factors ('Kennecot Journey' 49).

Brosseau's work is of importance to this study for crystallizing a methodological intervention – the championing of close reading and recognition of the complicating factors of textuality in the production of 'another type of geography' and resultant accounts of place – but also, perhaps more centrally, for suggesting the reformulation of the relationship between experience and representation for literary geography. Since Brosseau wrote this article, the experiential has become something of a determining category in cultural geographical research, but its compatibility with representation and its associated fields of enquiry has not always been recognized. It is towards this problem – which Brosseau holds in tension rather than resolves – that my own work is turned. In short, I aim to capitalize on Yi-Fu Tuan's observation – often cited but seldom fully taken account of – that 'a function of literary art is to give visibility to intimate experiences, including those of places' (162).

**The Challenges and Opportunities of Non-Representational Theory**

This monograph is concerned with the representation of place in contemporary writing. More specifically, it is concerned with the representation of experience of place. One of its central claims is that (in the texts studied here) this second statement is constitutive of the more universal first: here, the representation of experience of place *is* the representation of place. This is the significant factor unifying a superficially disparate set of writers and informing the claim I am implicitly making, that their work collectively signifies a new or newly rediscovered poetics for literary landscapes. Julie Sanders notes that 'one of the more productive sites of encounter between geography, history, and literary criticism as scholarly enterprises has certainly been the return to "description" as a social art form, an aesthetic that each of these disciplines benefits from reconsidering in its multiple manifestations and formations, not least across time' (294). This study is centrally concerned with what might be called the descriptive aesthetics, alongside the narrative forms, of writing on landscape and place.

The pairing of representation and experience indicates some of the theoretical concerns of the monograph. Whilst, as noted above, interdisciplinary dialogue has taken place for a number of years and continues to do so with appreciable results, significant methodological differences remain. Exchange has commonly taken the form of correcting perceived imbalances in research methodologies and priorities, and it is in this shape that non-representational theory has come to inform a significant body of work in cultural geography. It is certainly in these terms that Ogborn conceives of the shift towards enactment in geographical research agendas: 'This turn to embodied practice is partly a reaction to the overdetermined readings of spaces and texts produced through notions of representation.' Thus, '[t]he idea of non-representational theory, as developed within Geography, wishes to rethink spaces and their making beyond the notion of representation in text and image' (148). A similar narrative appears in Ben Anderson and Paul Harrison's account of the ascendency of non-representational theory in human geography. Social constructivism, which preceded its advent, was 'distinguished by a preoccupation with representation' and took 'the collective symbolic order' for its 'primary ontological object'. 'Epistemologically', they continue, 'this means that the "action" is *not* in the bodies, habits, practices of the individual or the collective (and even less in the surroundings), but rather in the ideas and meanings cited by and projected onto those bodies, habits, practices and behaviours (and surroundings).' From this perspective, 'there can be no sense of how meanings and values may emerge *from* practices and events in the world' (4, 5, 6). It is this oversight that non-representational theory looks to correct. It emphasizes the practices and experiences that go into the making of representations but are often lost in scholars' treatments of the latter. As Harrison argues elsewhere, 'it is only through stalling the working out of our traditional modes of explanation … that we may acknowledge the implications of the performative and, in this acknowledgement, learn how to see and, more importantly, to *show* practice as that which incessantly, irretrievably, excessively, *happens* – as that which is *taking-place*' (489–90). Whilst these critiques are concerned primarily with trends in geography, the shift identified can be seen as taking place across a much wider range of academic fields. Tim Ingold charts a concurrent development in anthropological research, where advocates of 'practice theory' argue that 'cultural knowledge, rather than being imported into the settings of practical activity, is constituted within these settings through the development of specific dispositions and sensibilities that lead people to orient themselves in relation to their environment and to attend to its features in the particular ways that they do' (*Perception of the environment* 153).

John Wylie has understood this trend in similar terms, finding that recent developments in cultural geography have been characterized by 'a rhetorical and substantive shift, from studies of *representations of* landscape, nature, identity, space, place, the body and so on, to studies instead investigating various *performances* and performativities of these tropes' (*Landscape* 163). In particular, Wylie suggests that 'currents of unease' circulated as a result of the powerful paradigms established by cultural geographers such as Dennis Cosgrove and Stephen Daniels, whose early work focused on the representational and symbolic aspects of landscape. For Wylie, this emphasis 'tended to elide the *materiality* of the landscape itself and also the … relations enacted through and within landscapes' (96–7). Cosgrove and Daniels's *The Iconography of Landscape* (1988) opens with the assertion that 'a landscape is a cultural image, a pictorial way of representing, structuring or symbolizing surroundings' and aims to explicate 'the status of landscape as image and symbol' (1).[3] Subsequent critical emphasis has fallen on pictorial traditions and those critiquing representational models have, like Ogborn above, tended not to differentiate between 'representation in text and image'. I do not deal extensively with the proposed limits of the pictorial, but rather stress the opportunities of textual poetics.

The aim of this monograph is to not sidestep the challenge of non-representational theory and the turn towards embodiment, but to profit by these methodologies and integrate their findings with due attention to textuality. Indeed, my work might be situated in relation to the integrative approach that is now characterizing current exchanges. Daniels et al., in their introduction to *Envisioning Landscapes, Making Worlds*, make clear that this volume's title 'describes two *reciprocal* aspects of human geography: envisioning landscapes – imaginative, reflective and representational; making worlds, ethical, performative and material. *These aspects dovetail and interact*' (xxvi, my italics). Within this reciprocation, landscapes are 'understood as experienced as well as envisaged, as lived in as well as looked at' (xxvi). In this context, I endorse Ogborn's assertion that recent shifts in cultural geography 'need not necessarily send a shiver down the spine of those concerned with the words on the page', and resist the notion that

[3] Daniels's later work, by contrast, has been central to cultural geography's engagement with performance studies. See Daniels, Pearson, and Roms's 'Editorial'.

non-representational theory and literary criticism are incommensurate (148). The dynamics of literary form recuperate experience.[4]

Nigel Thrift, outlining 'the main tenets of non-representational thinking' whilst also respecting its diversity, emphasizes a shared suspicion of prevailing representational models:

> These schools of thought all deny the efficacy of representational models of the world, whose main focus is the 'internal', and whose basic terms or objects are symbolic representations, and are instead committed to non-representational models of the world, in which the focus is 'external', and in which basic terms and objects are forged in a manifold of actions and interactions. (6)

Thrift's summary goes on to note that non-representational theory, deriving from a philosophical lineage running through Wittgenstein, Heidegger, and Merleau-Ponty, finds that 'practices constitute our sense of the real', that 'the particular moment' is valorized, as is 'thinking with the entire body' as opposed to the focalization of the visual and the marginalization of the aural and haptic that apparently accompanies representational models (7).[5] However, these tenets are not 'to deny processes of cognition, or the reality of representations', but rather to 'situate these imagined understandings as only a part of a broader process of knowledging' (Thrift 8). This thinking has undoubtedly been important in critiquing form in representation, and registers as such in the texts that this study approaches. (For example, the treatment of mapping in Chapter 2 finds cartography, in its official and military manifestations, to be a mode of representation that occludes practice in problematic ways.) For the writers studied at length here, this critique has acted as a spur to find new ways of representing place, and, crucially, to employ new textual strategies in doing so. What is happening, as a result, is a re-investigation of the relationship between – or perhaps even a coalescence of – representation and experience, along with a textual recovery of the fullness of the 'knowledging' process that Thrift identifies.

If the trajectory of non-representational theory has been not to abandon representation *per se* but to restructure the emphases of study in favour of practice and sensory engagement, some of its practitioners have tended to hold representation and experience apart and install an oppositional relationship between them. Recent developments in the field have attempted to assess the permeability of this boundary. Hayden Lorimer has retitled the movement as 'more-than-representational' geography with an 'extrarepresentational form of address'

---

4 Again, anthropological research provides a pointer here: Victor Turner's work finds that experience is itself 'instinct with form'. It is with this in mind that I suggest that the techniques of textual representation *recuperate* that which is already present in the experience in the first place (*From Ritual to Theatre* 13).

5 The important texts include: Ludwig Wittgenstein, *Philosophical Investigations*; Martin Heidegger, *Being and Time*; Heidegger, 'Building Dwelling Thinking'; Maurice Merleau-Ponty, *Phenomenology of Perception*.

because 'the teleology of the original "non-" title [has] proven an unfortunate hindrance' (84, 86). However, the recommended shift in focus and the retention of an idea of representation as 'rendering inert all that ought to be most lively' still suggests a well-policed border (84–5). David Crouch notes that 'representations are the product of living and emerge through practices', and that thinking of the two relationally might 'deepen an articulation of landscape in relation to life, and space in relation to living' (7). However, even here, the establishment of a relationship still tends to award primacy to anterior experience rather than representation. It remains the case that whilst the separation of representation and experience may not always be a theoretical aim, it has often informed findings in practice. The reluctance to engage with non-representational theory on the part of literary critics no doubt results from perceiving it in these terms. Whilst critical discourse has sometimes been caught up in this separation, recent and contemporary creative writers are elaborating composite approaches to place that navigate and transgress disciplinary boundaries. This composite approach allows for the combination and interplay of experiential and symbolic registers of landscape in light of what Wylie describes as 'the complex, reciprocal relations' between 'discursive practices' and 'histories and traditions' ('Becoming-icy' 251).

To varying degrees, notions of walking inform all of the texts with which this monograph engages, and it is thus with recourse to theorizations of walking that the relationship between representation and experience might best be described here. Thrift's most fundamental summation of non-representational theory rejects the privileging of the 'internal', associated with symbolic representations, and favours instead the 'external' realm of actions and practices. Yet for A.R. Ammons, a walk is 'an externalisation of an interior seeking, so that the analogy is first of all *between the external and the internal*' (116, my italics). This relationship informs literature about walking, landscape, and place. Roger Gilbert, describing what he calls the 'walk poem', finds it to 'consist in a desire to erase the difference between text and experience, to assert and sustain an absolute coincidence of language and bodily sensation'. The walk poem, he observes, 'takes as its subject not a landscape, an essentially static entity, but an experience. It is thus *transcriptive* rather than descriptive' (4, 8). He continues:

> A kind of fluid oscillation between external objects and inward ideas may ensue, one whose very rapidity blurs the borderline between physical and mental experience. … It is in the intricate dialectic between perception and reflection that the walk poem finds its centre … Thinking is shown to be not a mastering of experience but a product of it, circumscribed by the temporal and spatial limitations of the body itself. (11)

The final sentence of this passage provides the basis for an answer to those who fear that representation obliterates practice in its presentation of a generalized, fixed, and framed picture of landscape (Cresswell, 'Landscape and the obliteration of practice'). For Gilbert, '[t]he walk does not transcend experience', but 'remains caught up' in it (18). Gilbert's reading tallies with Anderson and

Harrison's suggestion that representations are apprehended as 'performative presentations' (19). Representation and experience need not be held apart as they appear to be in some versions of non-representational theory, but rather the recognition of their complexes of interaction can inform a study of literary landscapes. Gilbert's terms ('fluid oscillation', 'intricate dialectic') suggest just such a dialogue.

However, my own study modifies Gilbert's theorization in noting that an oscillation implies a two-way movement. Gilbert's acknowledgement that thinking and representing are conditioned by the spatio-temporal particulars of experience is undoubtedly justified and has a lot in common with Thrift's idea of 'context'. For Thrift, epistemology is informed by the context of embodiment: 'a practical or situated way of knowing is contextual, and rooted especially in embodiment' (33). Context is conceived of as 'a performative social situation' that is more or less spatially extensive and more or less temporally specific' (41). My contention here is that thinking of the context of experience 'especially' in relation to embodiment in practice too often becomes thinking of context *only* in relation to embodiment; and that defining context as 'more or less' spatially and temporally delimited tends towards the more specific end of the spectrum. In short, critical approaches to landscape informed by non-representational theory, in an attempt to zero in on bodily engagement and to focus on practised experience, have tended to neglect other kinds of context in which the event of encountering place might be conceived. Historicity has been substituted for immediacy, embedded knowledge for perception. The counteractive to solely representational accounts of landscape has over-corrected the error to neglect in turn the role that inherited traditions and a history of representation have to play in conditioning experience. I contest that non-representational theory and the phenomenological accounts of place that it has produced have sometimes neglected the narratives in which events occur, and that a prior history of representation which is crucial to a full account of place has been missing from these formulations. In response, this study matches an examination of the ways in which the circumstances of practice condition representations produced out of them, with a coterminous examination of the ways in which writers' and readers' knowledge of other representations condition that experience. I wish to complicate the notion that experience informs representation in a unidirectional manner, and to demonstrate the extent to which the former *precedes* as well as *proceeds from* the latter.

## Articulating, and Other Terminology

'Articulating place' goes some way towards summing up this synthesis. The verb 'articulate', in its range of applications, suggests something of both sides of the debate.[6] 'To express distinctly; to pronounce distinctly; to utter, give utterance to. Also: to express in words, esp. clearly and fluently': in short, to represent, orally or,

6 All definitions given henceforth are adapted from the *Oxford English Dictionary*.

by extension, textually. 'To join or unite; to attach or unite (esp. a bone) by a joint': pertaining to the body, to the experience and practice of movement. Two examples from recent poetry illustrate something of the meeting of different meanings that 'articulation' is used to suggest. Thomas A. Clark's *The Hundred Thousand Places* (2009) is a book-length poem describing a long walk (or series of walks) through an environment encompassing a range of coastal and mountainous landscapes. It conforms in many respects to Gilbert's idea of the 'walk poem', especially in its marked 'desire to erase the difference between text and experience' (4). To this end, the poem is written entirely in the second person: Clark's eschewal of first-person reportage can be read as an attempt to efface (or at least to interrogate) the subjective projection of meaning into or onto place, and to emphasize instead the processual encounter. To the same end, the poem's short lines and clipped forms set limits on expression, offering description more often than interpretation, and focusing on the way in which 'you feel your way' through the landscape's 'distances' and 'volumes' (50). Geometric vocabulary here emphasizes immersion. Nonetheless, historicity plays a key role:

> a common idiom
> carries through
> complex articulations
> call it a place
>
> it was not your
> intention to bring
> all your resources
> here but you do.
> (69)

Despite a primary focus on the feel of the walk and the experience of the senses therein, the passage notes the 'common' (shared, inherited, learnt, and customary) ways of thinking and writing about (that is, 'articulating') landscape which all enter into the 'complex' of producing an experience of place in the immediate encounter. Furthermore, this presence of prior representations is unavoidable: irrespective of 'intention', all of these cultural 'resources' are present in the moment of encounter. What we bring to a place is crucial: it conditions our practice and experience there. The poem recognizes and recuperates the interaction of representation and experience.

John Burnside's poetry is often focused on issues of dwelling and invites a phenomenological reading of place. In its emphasis on articulation in terms of the body's movement, '*De Humani Corporis Fabrica*', from the collection *The Good Neighbour* (2005), can be interpreted in this way.

> I've read how Michelangelo would buy
> a stolen corpse
>                     to study
>                                  in the dark

> the movement of a joint
> or how a face
> articulates the workings of the heart.
> (6)

However, in this instance, 'articulation' offers a meeting point of the exterior and the interior (the face and the heart). Additionally, Burnside is not unaware of the *textual* resonances of the term – importantly, though the poem seems to set bodily experience against cerebral activity, it nonetheless opens with an account of reading Vesalius's sixteenth-century anatomical study, *Fabrica* (whose title it replicates in its own). Additionally, in the section quoted here, it is noteworthy that Michelangelo – a painter engaged in representing the body's articulations – is the focus. The presence of book-learning, though not magnified, cannot simply be overlooked: representation is already implicated in understanding anatomy and experience. Furthermore, the poem itself is, in some sense, articulated – the layout of text on the page turns around line-breaks that act as joints or hinges:

> but I have never learned
> nor wished to learn
> how bodies work
> other than when they move
> and breathe.
> (6)

Furthermore, textual articulation is itself related to the body here: read aloud, the hinged lines impose a breathing pattern on the reader. Just as the poem's form and subject matter interact in this passage, representational strategies and embodied experiences also meet here.

For both poets, the production of place is specific and complex, and ties in with additional meanings of the verb 'articulate'. 'To formulate in an article or articles; to set out in articles; to particularize, specify': suggesting the process of overdetermination whereby places are produced from a number of inputs (including those drawn from sensory perception *and* a history of representation), and the particularity with which each place and each experience of place is constructed. Finally, 'To meet at or form a joint (esp. with another bone); to be united by a joint': articulations of place might be considered the meeting or joining of representation and experience. Crucial to the critical recognition of this meeting point is a literary geography that pays attention to the formal properties of its objects of study. To this end, the concept of articulation that I have elaborated here is closely aligned to Thacker's commitment to 'understand how social spaces dialogically help fashion the literary *forms* of the … text'. 'Literary texts', Thacker goes on to note, 'combine both metaphorical and material space if we focus upon their formal properties.' The interaction is reciprocal: '[l]iterary texts represent social spaces, but social space shapes literary forms' (*Moving Through Modernity* 4, 28). Here, I trace the ways in which forms of experiencing place are re-presented in the forms of literary texts.

The monograph title defines its field of study as 'contemporary literary landscapes'. This description – along with the term 'landscape writing' that I have also used here – is selected deliberately for its generic openness in demarcating the kind of writing studied. Wylie has noted that 'whereas landscape art is a familiar and rich visual tradition, there is in contrast no generally understood or accepted literary genre named "landscape writing"' ('Writing through Landscape' 48). This study can be considered a step towards assessing the characteristics that this term might denote. The analyses that follow range over verse and prose, fictional and non-fictional samples without introducing a hierarchy amongst them. It is one of the contentions of my work that a recently emerging mode of landscape writing necessarily shuttles across generic boundaries: this mode of representation reflects the fact that the landscapes traversed and described are built out of a complex of real and imaginary sources. A further point of methodology is made in the emphasis placed on the value of the critical apparatus of literary criticism in the study of '*literary* landscapes'. My work draws on cultural geography but, I hope, also gives back new insights in analyses that foreground the place of textuality. My readings of texts, in contextualizing the articulation of landscape within a history of representations, accentuate the important role played by narrative in the formulation of place. What I am suggesting here builds on Cronon's description of environmental history as 'tell[ing]' landscape 'stories' ('Kennecot Journey' 33).

The monograph speaks of place more often than space. The difference and interaction occurring between space and place is a topic with a long critical history. The connotations of 'place' are given by Yi-Fu Tuan, who describes places as 'centers of felt value', and finds that '[w]hat begins as undifferentiated space becomes place as we get to know it better and endow it with value' (4, 6); and by Tim Cresswell, who suggests that a place is 'a meaningful location', and goes on to note that '[a]s well as being located and having a material visual form, places must have some relationship to humans, and the human capacity to produce and consume meaning' (*Place* 7). Here, 'place' is selected for its suggestion of heightened specificity and particularity to be found in the case studies that follow. If the poetics by which these texts proceed is one that articulates and specifies, the sites that are described tend towards the humanized (that is, experienced) qualities of place over and above those of abstract space. From another perspective, Michel de Certeau's assertion that 'space is a practiced place' has led to some favouring of the former term (117). However, as Eric Falci notes, after the spatial turn, '[t]he assumed immobility of place was infused with the indeterminacy and activity of space. … Places – actual, virtual, conceptual – became the site of spatial practices and experiences' (201). In short, 'place' has taken on the positive attributes of 'space'. As a result, positions such as that of J.E. Malpas emerge: 'place is integral to the very structure and possibility of experience' (31–2). Here, then, 'place' is taken to denote the specificity arrived at out of experience, without eliding the process whereby it is achieved.

'Landscape' is defined as an area of focus for its associations with both an area of land and a tradition of representing place (the doubleness of the word is explored at greater length in Chapter 1), and the inherent suggestion of a *constructed* environment. Related to this, 'articulating', 'picturing', 'mapping', 'walking', 'engaging', and 'noticing' in the section headings are given as present participles to indicate the ongoing, open production of place. The constant and reciprocal interaction of representation and experience suggest place not only as location, but also as process. Here, my work is indebted to Henri Lefebvre's instigation of a shift towards thinking of space and place as process in his proposal of 'trac[ing] the coming-into-being and disappearance of [their] codings/decodings' (18), and to Doreen Massey's insistence that 'we recognise space as always under construction … always in the process of being made' (*For Space* 9). Furthermore, the central argument of the monograph, that these textual representations signify not fixed versions of landscapes but records of particular experiences of those landscapes, carries the implication that they do not close the process, but rather that they themselves enter into the very traditions and discourses of representation on which they draw, thus feeding into ongoing interventions (Lefebvre 41–2).

Just as places are in process rather than closed or completed, modern literary theory proposes that texts do not present singular, stable meanings but are instead ambiguous and can be interpreted variously. Hones notes that for literary geography these two observations might be associated: 'This view of text relates well to an understanding of place as something relational, unfinished, and dynamic, internally various and the product of interconnected human and non-human histories' (1310). In the first instance, this dual indeterminacy signifies an opening for landscape writing and suggests a procedural opportunity. However, it also poses a dilemma and prompts questions: does describing a place's meaning signify imposition? How readily does the description of a personal experience of place shade into ascribing a set reading of that place? How is the balance between what Hones calls 'human and non-human histories' to be configured? In their shared combinatory aesthetic linking representation and experience, the writers with which this monograph deals all address this set of concerns (though with varying outcomes), and all write markedly subjective accounts of place. It is not my intention to elaborate conclusions for these issues here, but to intimate that attention to these concerns informs the analyses that follow. Building from this, the book's conclusion groups and consolidates findings in relation to these questions.

## The Sample, and Chapter Summaries

This study comprises five extended case studies, each focusing on selected texts by a single author. It does not aim to offer an exhaustive appraisal of contemporary writing of landscape and environment, but rather selects facilitating texts for analyses that both crystallize more general trends in the current literary treatment of place and address the theoretical and methodological concerns that I have described. As such, this study is attentive to one of the main challenges facing literary geography and the attempt to establish cross-disciplinary dialogues. Hones suggests that 'work in the field has conventionally been sorted and labelled in such a way as to indicate a primary authorial audience defined by its interest in a shared topic or text, and only after that in terms of an interest in literary geography as such'. Consequently, '[t]he potential development of a collaborative community in literary geography has been seriously inhibited' and 'so much that is shared and important' has been obscured (1312–3). My own work, ranging across texts, authors, and genres, responds to this situation. It differs from the implications of Hones's summary in employing a methodology that looks to make broad assessments of literary geography *through* close reading of particular texts (rather than holding the two tasks apart). Just as this work does not attempt to survey the whole field, even within the oeuvres of those authors receiving extended attention facilitating texts are dwelt upon to address pervasive thematic and aesthetic concerns.

Indeed, it is in the poetics deployed in the description of place that links between the texts studied emerge. All of the writers approached are acutely aware of the role played by literary form in landscape writing: the spatial text and the space of the text are intricately related in their work. It is noteworthy, in this respect, that in various ways all of these authors produce texts that question form and genre. W.G. Sebald's prose narratives are located on the margins of an array of different literary forms and draw on a wide range of traditions; Iain Sinclair and Ciaran Carson have both published poetic, fictional, and non-fictional work, as well as texts that combine these forms or undermine the distinction between them; Robert Macfarlane and Kathleen Jamie are centrally concerned to interrogate a set of generic characteristics that embody nature writing.

Whilst critical consensus regarding the importance of particular writers in the contemporary field is always emerging rather than achieved, the writers addressed here have, for the most part, had significant scholarly attention conferred upon them. Furthermore, they have often been cited as paradigmatic examples of the literature of place: Urszula Terentowicz-Fotyga finds Sebald and Sinclair to be exemplary instances of a 'new urban imaginary' (307);[7] Jonathan Bate credits the same pair (along with Roger Deakin, who has been extremely influential in Macfarlane's work) with leading the way in the recent rejuvenation of the English journey narrative and 'shap[ing] the prose of a new generation of non-fiction writers' ('To the River' n.p.); Eric Falci dwells at length on Carson in an article selecting only three writers through which to demonstrate recent developments in the poetic treatment of landscape, space, and place. Sebald and Sinclair in particular have attracted critical responses from literary critics and cultural geographers, making their texts a sound basis from which to make theoretical and methodological comparisons. Robert Macfarlane and Kathleen Jamie are leading exponents of the new developments in nature writing that demand but are only now beginning to receive scholarly attention.

If it is apparent that these writers are important to the field of landscape writing singularly, there is also abundant evidence to link their different projects

[7] Terentowicz-Fotyga's focus on Sebald as an urban writer emerges out of her close study of his text *Austerlitz*, whereas my own work is concentrated more centrally on *The Rings of Saturn* and more often relates his work to rural traditions.

to one another. *The Re-Enchantment*, a national project funded by Arts Council England that has sought 'to interrogate the various meanings of "place" in the twenty-first century', is paradigmatic of the way a number of these writers have been brought together. One of its core outputs in 2010 was a volume of essays in which writers respond to particular places, and to which Sinclair, Macfarlane, and Jamie were amongst the contributors. *The Re-Enchantment* also commissioned Grant Gee's film *Patience (After Sebald)*, a retracing of Sebald's East Anglian walk and including interview material from Sinclair and Macfarlane testifying to his influence on their thinking. Elsewhere, Carson has also suggested Sebald has been an influence on his own work (qtd. in Kennedy-Andrews, 'For all I know' 16). Whilst the group of writers addressed here do not signify a movement in any programmatic way, they are explicitly aware of the relations between their writing. The potential significance and critical possibilities that these interconnections make for in this project are the identification of a common approach, perhaps even a prevailing trend, to descriptive aesthetics and to narrative form across a range of contemporary landscape writing. The category of intertextuality identified is not one of direct allusion to each other's work, but a shared awareness of modes of writing – a *bricolage* of practice – in the representation of place.

Chapter 1 finds that W.G. Sebald's genre-defying prose narratives offer a contact zone between different academic disciplines and the opportunity for overcoming the differences identified in this introduction. Sebald's texts read the ubiquitous historical narratives found in topography. *The Rings of Saturn* is a semi-fictionalized account of a walking tour of coastal Suffolk that finds history's traces everywhere. The landscape, though experienced first-hand, has been described so much that this experience is mediated through other people's eyes. Because this mediation is a significant part of the experience, Sebald's writing is extremely attentive to processes of picturing. This chapter opens with an investigation into the varying resonances of the term 'landscape', comparing its painterly associations with framed and fixed views of places as scenery to a more dynamic and processual notion of material spaces and practices. Sebald's writing draws out uncertainties that inhabit 'landscape' and capitalizes on the traffic between its multiple valences.

Chapter 2 charts the development of Ciaran Carson's poetry of place and his spatial poetics in response to the Northern Irish Troubles and the sectarian geographies that they perpetuated. Carson's oeuvre offers a sustained focus upon the cityscape of Belfast from *The Irish for No* and *Belfast Confetti* to more recent figurations in poetry and prose such as *The Star Factory*, *Breaking News*, and *For All We Know*. These texts engage extensively and specifically with the materiality of Belfast, and develop a poetics reflecting on ways of seeing. In contrast to the landscape of checkpoints, Carson interrogates mapping processes to reveal the multitude of perspectives present in a polyvalent and indeterminate city. If the authority of the map is discredited, this chapter argues that the labyrinth provides a more dynamic alternative. This motif, both as a conception of the city's fabric and as a figure of narrative structure, constantly undermines positions of certainty and the all-encompassing, definitive perspectives. The polyphony and maze-like qualities of Carson's poetic storytelling make it fertile ground for resistance to the imperatives of the Troubles: in its intricacy and proliferation, narrative necessarily exceeds and therefore contests the entrenched positions of the sectarian conflict. To this end, Carson develops a labyrinthine, woven aesthetics of the space of the text itself.

Iain Sinclair's texts are often critiqued for their strong reading of place but Chapter 3 argues that his later writing loosens narrative bounds to find a representational form that navigates between exploration and exploitation, and is capable of dramatizing the experience of place. I focus on *London Orbital*, Sinclair's circumnavigation of London loosely following the M25 motorway, along with *Lights Out for the Territory* and *The Edge of the Orison*, with which it forms a loose trilogy. Walking is the pre-eminent generative process at Sinclair's disposal and thus receives extended treatment in these three texts. Whilst many critics find Sinclair's city to be a textual palimpsest overlooking embodied practices, I suggest that he utilizes the space of the text to recuperate ambulatory perspectives and perceptions. If the rhythm and pace of walking facilitate a certain kind of engagement with place and produce a particular kind of text, the relationship is reciprocal: Sinclair's paratactic style aims, structurally and syntactically, to reproduce the experience of walking in textual form.

Robert Macfarlane's new formulations of nature writing are defined by the exploration of humans' relationship with the non-human environments they inhabit. This often takes place through the employment of hybrid, generically ambiguous literary forms. Chapter 4 focuses on the testimonial approach to engaging with wildness, nature, and place that is evident in his evocative texts: *Mountains of the Mind*, *The Wild Places*, and *The Old Ways*. I argue that each of Macfarlane's texts is characterized by a narrative arc that dramatizes a trajectory away from one mode of conceiving of place's meaning and towards another. They reproduce this change in the structure and form of the text, employing novelistic techniques of deferral and revelation. The notion of 'wild' shifts from landscape untouched by human intervention towards one in which human and non-human influences enter into complex, fluid interactions. Wild places are no longer seen as spaces without history, but conversely as sites in which histories remain traceable.

If Macfarlane's texts are testimonial, Kathleen Jamie's writing adopts a more documentary tone focused on attentive recording. She offers differently accented refractions of a similar set of concerns and critiques habitual ways of framing nature. Chapter 5 examines her prose texts, *Findings* and *Sightlines*, and her poetry, particularly *The Tree House* and *The Overhaul*. Jamie shifts the focus of nature writing towards issues of observing, noticing, and paying attention. Jamie's prose style is characterized by doubling back on previous observations and assertions, often in short, clipped phrases that, collectively, create an aesthetic of reticence in the description of nature. On a larger scale, her episodic texts consciously resist the temptation to formulate grander narratives. Jamie is attentive to nature, but also aware of the limitations against which that attentiveness presses.

The terms with which each case study is associated – Sebald with 'picturing', Carson with 'mapping', Sinclair with 'walking', Macfarlane with 'engaging', Jamie with 'noticing' – does not indicate a relevance of each of these terms exclusively to that writer. Indeed, the opposite is true. For example, whilst walking is treated in detail in relation to Sinclair's texts, Carson's texts also embody, as Neal Alexander suggests, a 'distinctively peripatetic aesthetic' (*Ciaran Carson* 85). Terms and writers are matched as seems most apposite to allow for theoretical and applied discussion, and for the development of dialogue between these two critical tasks. That each writer thinks in a variety of terms about articulating place is perhaps the defining hallmark of the composite approach that I describe above.

Flowing out of the final chapter's approach to the problem of imposing meaning on place, the afterword argues that merely transcribing place – giving an objective account – is problematic in literary landscapes. The extent to which an aesthetic privileging the experiential approximates transcription (and whether or not this is its aim) comes under scrutiny here. In its selection of texts, this monograph confers attention on writers who emphasize rather than elide the subjectivity of their responses to place. Sebald, Carson, Sinclair, Macfarlane, and Jamie do not write of place *per se* but of an experience of place in a sensuously responsive mode, marking their subjective rendering of place not to claim authority but to deflate any claim to a definitive vision. These texts bear witness to experiences of place. The immersion that occurs there is physical and textual, embodied and representational, and, above all, partial. In response, contemporary landscape writing is partial in another sense: it is generically ambiguous and moves across formal divisions. The monograph thus concludes by noting how landscape writing offers a paradigm for this shift, perhaps the most important occurring in literature today.

# Chapter 1
# Picturing: W.G. Sebald

This chapter opens with a traveller's account of a landscape. When he visited the Lake District in 1855, Nathaniel Hawthorne wrote:

> On the rudest surface of the English earth, there is seen the effect of centuries of civilization, so that you do not quite get at naked Nature anywhere. And then every point of beauty is so well known, and has been described so much, that one must needs look through other people's eyes, and feel as if he were seeing a picture rather than looking at a reality. (Qtd. in Drabble 7)

This passage formulates key issues for consideration in a reading of W.G. Sebald's writing on landscape. Sebald shares with Hawthorne a mode of engaging with place that reads historical narratives in topography. Indeed, for both writers, history is ubiquitous even where its marks might be supposed to be least obviously present. Just as Hawthorne finds 'the effect of centuries of civilization' to be unavoidable even in the 'rudest' (that is, most rural and least industrialized or perhaps even least modern) region, Sebald's *The Rings of Saturn*, a semi-fictionalized account of a walking tour of coastal Suffolk, finds that history's 'traces' are everywhere 'even in that remote place' (3). The second comment that Hawthorne makes also resonates in Sebald's landscape writing. The landscape, though experienced first-hand and for the first time, has been 'described so much' that this experience is mediated 'through other people's eyes'. This mediation is, in fact, a substantial part of the experience. Glossing this passage from Hawthorne, Margaret Drabble finds that it is impossible not to 'see certain landscapes through the eyes of the writers that discovered them', and suggests that '[k]nowingly or unknowingly, we have all been influenced by the writers who went before us' (7). Though the notion of writers 'discovering' landscapes requires interrogation, the unavoidability of reference to previous artistic visitation certainly plays a large part in Sebald's work: his writing of the experience of place is extremely attentive to this process of picturing. Whilst Drabble silently converts the pictorial tradition that Hawthorne actually refers to ('a *picture* rather than a reality') into a literary one ('the *writers* who went before us'), Sebald's texts, I argue here, are invested in unpacking the relationship between the painterly and the textual, as well as the role of the visual in both. 'Landscape' is a term freighted with these representational connotations. Its investments, limits, and possibilities require some preliminary commentary here.

The role played by prior representations in the formulation of landscape has often been the issue on which interdisciplinary dialogue in this area has foundered. Simon Schama, in *Landscape and Memory*, found that '[b]efore it can ever be a repose for the senses, landscape is a work of the mind. Its scenery is built up as much from strata of memory as from layers of rock' (6–7). Allowance is made for a degree of mutuality between memory and materiality in the determination of place here, but after the turn towards performance in landscape studies the primacy awarded to anterior representations in this vein has become problematic. Interest in embodiment has led many to demur from statements, such as Schama's, affirming that 'it is culture, convention, and cognition' that 'invests a retinal impression' with the qualities we ascribe to a landscape (12). The same problems might be said to persist in more recent constructions. For example, Bertrand Westphal, discussing what he calls 'the saturation of a place by the text', writes that literature 'anticipate[s] geography', and that '[t]he text precedes the place, and sometimes seems to anticipate its discovery' (154–5, 158). For some, this anteriority restricts present experience.

In 'Landscape and the Obliteration of Practice', Tim Cresswell identifies some of the problematic aspects of this idea of landscape for contemporary cultural geographers. The term 'does not have much space for temporality, for movement and flux and mundane practice. It is too much about the already accomplished and not enough about the processes of everyday life' (269). The titular emphasis of 'landscape' in the present book is a conscious effort to rehabilitate the term in the face of charges such as Cresswell's, and is indicative of a desire to suggest the performativity of representation. For Cresswell, the gulf between landscape and practice is at the heart of the problem:

> Landscape, on the one hand, appears to encapsulate the notion of fixity – of a text already written – of the production of meaning and the creation of dominating power. Landscape is solid. Practice, on the other hand, is about fluidity, flow and repetition. It is about the negotiation between continuity and change. (270)

To overcome this impasse, Cresswell suggests, we need to establish dialogues between the two terms and think of vision, associated primarily with landscape, as a practice. 'Landscape' can thus be thought of as 'a practised environment', and is subject to 'an injection of temporality and movement into the static' (277, 280). Others have also called for a rethinking of these terms. David Crouch has suggested that 'landscape and space might be conceptualized relationally', that is, landscape theory might profit from conceptions of space as process, thus opening up 'possible ways in which landscape can be recognized as dynamic and processual through the consideration of representations relationally with the character of life and its practices' (6, 14). Much of the impetus for these attempts to incorporate process into landscape has come from the opening of dialogues with performance studies. In particular, Mike Pearson's work has insisted that landscape 'is not something to be looked at but something to be lived in', a revision accompanied by a concomitant 'shift from the optic to the haptic in the apprehension of landscape' (11). Moreover, in *Process: Landscape and Text*, Catherine Brace and Adeline Johns-Putra testify that process is now an integral part of configuring 'the relationship between our surroundings and our imaginings', and offers a juncture for the meeting of representational and non-representational models of interpreting landscape (20). In their introduction to a special issue of *Geoforum* entitled 'Enacting geographies', Dewsbury et al. record that for those engaging with non-representational theory, despite an avowed focus on 'processual registers of experience', representations are not to be discarded but instead 'apprehended as performative in themselves; as doings' (437–8). Conceived thus, representations of place are not fixed or reified.

The relationship between practice and text or image has also undergone revision in this refocusing of the term 'landscape'. Proceeding from the principle that literary texts 'are not transparent accounts of a sense of place', Mike Crang considers that '[l]iterary landscapes are best thought of as a combination of literature and landscape, not with literature as a separate lens or mirror reflecting or distorting an outside world' (46, 57). John Wylie's account of the configurations of self and landscape produced by walking, whilst focused primarily on the affects and percepts of sensory engagement, acknowledges that 'attunement' to a history of cultural meanings and sensibilities (including a literary 'romantic inheritance') 'remain[s] the precondition and the *milieu* of contemporary countryside walking in England'. As a result, walking in a landscape 'does not in any straightforward way constitute an "embodied" connection or immersion that is foreign or resistant to the knowledges produced by gazing, contemplating or navigating' ('A single day's walking' 235, 237, 240). In short, the turn towards considering landscape in relation to practice has not (for everyone) involved turning away from landscape as represented in text and image. Rather, it has led towards a recognition of the entwining of the two: the practice of producing representations of landscape, and the embedding of practical engagements with landscape within a history of representation. In the reading of Sebald to follow, the optic is not so much replaced with the haptic; but instead, the *process of picturing* – the combination of experience with references made to previous representations, and the inclusion of photographic pictures and other images in Sebald's texts – is dealt with at length.

Before his death in 2001, W.G. Sebald produced four extended prose narratives – *Vertigo* (originally published in German 1990, translated 1999), *The Emigrants* (1992, 1996), *The Rings of Saturn* (1995, 1998), and *Austerlitz* (2001, 2001) – alongside shorter pieces and an academic output at the University of East Anglia.[1] Amongst these texts, *The Rings of Saturn* is taken in this chapter as a platform from which to offer a critique of current understandings of landscape from a perspective that looks to assimilate the findings of cultural geography and literary criticism. Sebald's texts defy certain classification (encompassing aspects of fiction, history, biography, and travelogue), and employ a mixed modality on

---

1 This study focuses on the English translations of these texts produced by Michael Hulse and by Anthea Bell. This decision is justified by the fact that, as Mark McCulloh notes, Sebald 'participated in the process of translation as an advisor to his translators … he was keenly interested in the problems of translation, and served as the first director of the British Centre for Literary Translation' (7). McCulloh argues for two separate groups of texts rather than a hierarchical division between superior originals and inferior translations.

the page (supplementing text with photographs and images). These factors have combined to draw critical responses from a wide range of existing academic disciplines as well as constituting paradigmatic case studies in the initiation of others (memory studies being chief amongst this latter grouping). It is the aim of my reading to situate itself at the contact zone between two specific academic disciplines invested in reading Sebald's texts – between their spatial aspects as found by cultural geographers, and a focus upon history and historiography in literary critical responses. A composite mode of reading may point towards ways of overcoming the 'different critical literacies' to which the introduction to this book has already drawn attention (Philips and McCracken 9). Indeed, it also suggests ways in which an awareness of and literacy in ideas of textuality and narrative have been productively introduced to our understanding and theorization of landscape.

Wylie has written of the importance of Sebald's texts to contemporary cultural geography. Sebald's work, he claims, 'has been a specific source of inspiration for recent geographical engagements with landscape writing', and 'has come to stand as something of a model for contemporary cultural geographies of landscape' (*Landscape* 207). Wylie's book *Landscape*, in which these remarks occur, charts the development of cultural geography's engagement with landscape and concludes with a focus upon Sebald's writing as a field in which the future prospects of the discipline might be explored. In an approach to these unclassifiable texts, an opportunity 'to develop newly critical and creative means of expressing relationships between biography, history, culture and landscape' emerges (206). Whilst this discipline's interest in Sebald is predicated on his spatial representations, critical responses from those engaged in literary criticism have tended to offer a more sustained focus on renderings of history in the same texts. In the preface to their aptly titled collection of essays, *W.G. Sebald and the Writing of History* (2007), Anne Fuchs and J.J. Long find that Sebald's 'writings relay a *Zeitgeist*' of the modern age because they display his 'sensibility for the individual's traumatic experience of history' (7). *The Rings of Saturn* is certainly invested in recalling the history of the places it reports on and in formulating a particular historiography. The text is replete with seemingly arcane and obscure historical matter and proceeds in antiquarian and curatorial styles themselves loaded with medieval and early modern precedents. W.G. Sebald's writings therefore offer a juncture at which these overlapping claims, made on behalf of space and place on the one hand, and historiography on the other, may be examined.

## Situated Perspective

My own study draws these disciplines into dialogue. I aver that Sebald's texts describe a 'situated perspective' which reads the traces of various tropes of history and memory in the landscapes that they traverse, and thus interrogate the variant fortunes of personal and general historical narratives. As such, Sebald's work responds to the problematic 'fundamental opposition' of memory and history identified by Pierre Nora in his theorization of *lieux de mémoire* (sites of memory). Nora finds that we are now confronted 'with the brutal realization of the difference between real memory … and history, which is how our hopelessly forgetful societies, propelled by change, organize the past' (8). In Sebald's texts, 'memory', constructed in a complex of spatial *and* temporal location, provides the basis for a critique of the erasure of the individual in traditional historiography. Illustrations of the aesthetics by which Sebald's critique is undertaken will chiefly be drawn from his ambulatory record of coastal Suffolk, *The Rings of Saturn*.

This text is selected for extended analysis here as it speaks to these theoretical concerns most readily and comes closer than Sebald's other major texts – *Vertigo*, *The Emigrants*, *Austerlitz* – to describing a specific landscape. It is geographically circumscribed and regional in a way that they are not. Whilst all four texts are characterized by spatial motifs of voyaging, *The Rings of Saturn* reports on a single, multi-day walking tour that is convergent rather than dispersed. Of the prose narratives, Simon Cooke finds it to be 'the most easily recognisable as a travelogue' (138). Sebald's itinerary follows a border, an edge-space, between land and sea. As such, it is the first of a variety of 'edgelands' that this book approaches, though Sebald's account traces mutability rather than offering a reified sense of interstitial space. Sebald perhaps selects the particular landscape for his journey due to the instability of the Suffolk coast and the marshy areas of this low-lying county. The coastal erosion for which these locales are famous provides a familiar model for the historiography of decline that the text inaugurates. It is within these broad parameters, themselves inaugurating a value-laden environment, that Sebald constructs a reading of place to which the mediatory role played by artistic representation is intrinsic.

J.J. Long, noting that Sebald's texts are constantly 'concerned with questions of perspective', finds that a 'recurrent topos is the desire for a stable, epistemologically reliable vantage point from which the object – be it a landscape or a historical event – can be represented' (111). Whilst this desire might be said to inform these texts, it does so as an ideal possibility characterized chiefly by its remaining unattainable. It is the irreducibly multiple and irreconcilable historical tropes that form the tenable subject matter of Sebald's work. In the texts' terms of description, what I call 'situated perspective' opens an array of tools at Sebald's disposal whereby this historiographical critique might be undertaken. The figure who variously observes, takes part in, and reports upon history is, in definite ways, *spatialized* and his or her perspective precisely *situated*. In this context, 'situated' does not imply a perspective that is stable, empirical, or authoritative, but rather is used to indicate the adoption of one perspective amongst the myriad of potential perspectives available. 'Situated' here indicates the contingent intersections of time and place at any given point in Sebald's narrative. As such, being situated is, in more than one sense, taking up a position: it does not claim objectivity and fixity, but emphasizes its own subjectivity and fluidity. Neither is a situated perspective incompatible with the mobile perspective that also informs Sebald's wandering, often circuitous texts. That *The Rings of Saturn* reports on a walking tour is significant in this respect. Rebecca Solnit suggests that for the walker, 'landscape moves by as a gently modulated continuity' – it is at this juncture of the situated and the mobile that Sebald's writing sits (174). In *The Rings of Saturn*, landscape is conceived by Sebald as a repository of collective memory (as it had been, he is clearly aware, by medieval and early modern writers before him). The narrator's choice of a route through a landscape is thus an idiosyncratic curatorial act of selection. Numerous instances of 'situated perspective' take place in contradistinction to the also present omniscient position of narration, with the effect that the former undermines the claims to universal vision made by the latter. To this end, situated perspective plays off its own instability. What is especially significant here is that this interpenetration of narrative idioms takes place in space, or in a specific place. It is the potential for comparison between narratological perspectives that a spatial dimension provides and that these texts exploit in their historiographical 'turning'.

## Landscape as Supplement

Cueppens has written that 'artistic perspective', as it is presented in Sebald's texts, provides 'a unique mode of access … unattainable from any other viewpoint' (66). *The Rings of Saturn* exploits the potentials that this provides for landscape writing in sustained ways. Greg Bond finds the landscapes of this text to be 'imaginary' as 'they are presented in works of literature, but also because they derive from and lead into the world of the human imagination' (33). Wylie finds them to be 'both starkly real and shimmeringly dream-like'; whilst they are characterized by 'circumstantial specificity and solidity', they also encompass 'irruptions of the surreal and the phantasmagorical' ('Spectral landscapes' 173–4). These traits suggest something of the artistic prism through which the East Anglian landscape of the text and the accompanying photographs is refracted, or perhaps, of which it is constructed and curated. Sebald, in *The Rings of Saturn*, might be said to conceive of artistic representation of spaces, of landscapes, as a particular perspective, that is, as a series of spatialized positions of reportage.

The reconciliation of such facets of Sebald's writing to a phenomenological account of place grounded in practised experience is potentially problematic. Jessica Dubow has recently described Sebald's texts as embodying a 'negative phenomenology' – a 'perceptive receptivity to absences and opacity' ('Out of Place' n.p.). Any sense of the transparency of a landscape experience is clouded by layers of mediation. As a consequence of the focus on embedded histories, these texts are characterized by an 'anti-mimetic gesture' in which 'all places are somehow posthumous' and landscapes signify 'a ground in which experience cannot take root' ('Still-Life' 189). In short, for Dubow, Sebald's landscapes are made up of historical presences now ostensibly absent from the territory. This kind of reading certainly seems to run the risk of precluding other aspects of 'the broader process of knowledging' that Thrift and other non-representational theorists have looked to reassert (Thrift 8). The phenomenological terminology in which Dubow's findings are issued ('perceptive receptivity', 'perceptual capacity') might be said to go some way towards mitigating the seeming opposition between two ways of thinking about landscape and place here. The same is true of Wylie's identification of 'solidity' alongside the 'phantasmagorical'. Even with these qualifications, Sebald's texts pose difficult questions for the accommodation of experiential and representational models of landscape. It is, in part, for this reason that this case study is the first of the five that this book constitutes – here, questions that run through all five are inaugurated rather than exhaustively answered.

Sebald draws on particular uncertainties that inhabit the very concept of landscape itself. Several cultural geographers have noted and celebrated the slipperiness of the term: Stephen Daniels has called it 'an ambiguous synthesis' (2: 206); David Matless has found that it 'productively migrates through regimes of value sometimes held apart' ('Properties of Landscape' 231). The *Oxford English Dictionary* defines 'landscape' first as pictorial, and then as material. It can be 'a picture representing natural inland scenery', 'a view or prospect of natural inland scenery, such as can be taken in at a glance from one point of view', and 'a tract of land with its distinguishing characteristics and features'. The second of these three descriptions usefully emphasizes the experience of seeing, or perhaps picturing to oneself, that shuttles between the representational and material aspects of the first and third respectively. Genealogy is also instructive: the pictorial sense of the term is the older, dating from the seventeenth century, with a move towards the inclusion of material landform under the term, adding a usage entirely divorced from the pictorial only by the late nineteenth century. It is also significant in relation to a discussion of Sebald's East Anglian landscapes that the term originates in Dutch artistic traditions on which *The Rings of Saturn* draws.[2]

Wylie conceives landscape as a tension between something we are *in*, and something we look *at*. He asks whether the word describes:

> the mutual embeddedness and interconnectivity of self, body, knowledge and land – landscape as the world we live in, a constantly emergent perceptual and material milieu? Or is landscape better conceived in artistic and painterly terms as a specific cultural and historical genre, a set of visual strategies and devices for distancing and observing? (*Landscape* 12)

Sebald's texts are not concerned to answer these questions but rather to draw upon the dual registers that the word possesses. The two opposed definitions of landscape as material (implying perception and engagement) and as representational (implying distance and framing) that Wylie suggests are collapsed together in Sebald's renderings. It is through the history of previous artistic visitation, coupled with his own sensory involvement, that Sebald engages with and inhabits the landscape that he describes. In this crucial respect, 'landscape', in the hands of Sebald, acts in a way similar to the Derridean concept of the 'supplement'.

Derrida's notion of the 'supplementarity' provides purchase by which to recognize the interdependence of practice and representation in the production of

---

2 For further details on the etymology of the term, see John Brinckerhoff Jackson, *Discovering the Vernacular Landscape*.

landscape in Sebald's texts, and in a widely applicable understanding of the term's complexities. The idea pervades Derrida's many texts but the clearest exposition of its meaning is issued in the famous sections of *Of Grammatology* in which Derrida deconstructs Jean-Jacques Rousseau's concept of the relationship between speech and writing. Reading Sebald's texts, artistic representations of landscape, both pictorial and textual (though especially the latter), might be said to bear the same relationship to their material counterpart landscapes as writing does to speech for Derrida. Derrida reads Rousseau's ambivalence to writing, examining the way in which, for Rousseau, it 'operates as a power of death in the heart of living speech: a power all the more redoubtable because it opens as much as it threatens the possibility of the spoken word'. As a result, Rousseau 'valorizes and disqualifies writing at the same time'. He 'condemns writing as destruction of presence and as disease of speech. He rehabilitates it to the extent that it promises reappropriation of that which speech allowed itself to be dispossessed'. In short, writing is seen as a 'restoration, by a certain absence and by a calculated effacement, of presence disappointed of itself in speech' (141–2). Derrida goes on to note that the two sides of Rousseau's thought are intrinsically bound together and that 'the word *supplement* seems to account for the strange unity of these two gestures'. Furthermore, 'the concept of the supplement', and, I am suggesting, landscape itself, 'harbors within itself two significations whose cohabitation is as strange as it is necessary'. In the first place, 'the supplement adds itself, it is a surplus, a plenitude enriching another plenitude, the *fullest measure* of presence. It cumulates and accumulates presence. It is thus that art, *technè*, image, representation, convention, etc., come as supplements to nature and are rich with this entire cumulating function' (144). But also,

> the supplement supplements. It adds only to replace. It intervenes or insinuates itself *in-the-place-of*; if it fills, it is as if one fills a void. If it represents and makes an image, it is by the anterior default of a presence. Compensatory [*suppléant*] and vicarious, the supplement is an adjunct, a subaltern instance which *takes-(the)-place* [*tient-lieu*]. As substitute, it is not simply added to the positivity of presence, it produces no relief, its place in the structure is assigned by the mark of an emptiness. Somewhere, something can be filled up *of itself*, can accomplish itself, only by allowing itself to be filled through sign or proxy. The sign is always the supplement of the thing itself. (145)

In this second function, the supplement substitutes the sign (a representation) for the thing itself, implying an insufficiency in the original. 'Landscape' functions in the same way: it signifies both the actual place and its pictorial representation. Pictorial representations add to but also substitute for their material counterparts. Sebald's landscapes are founded as much on the former as they are on the latter and they constantly draw attention to the dialogic element of their construction. The practice of landscape that his texts depict is characterized not only by immediacy of presence, but also by the mediated adjoining of meaning and value accorded in a history of the artistic, literary, and cultural visitation and representation upon which Sebald draws heavily.

Furthermore, Derrida's conception of the supplement also finds that the originary presence cannot be worked back to. Rather, it is the effect of an unending 'sequence of supplements': 'an infinite chain, ineluctably multiplying the supplementary mediations that produce the sense of the very thing they defer: the mirage of the thing itself, of immediate presence, of originary perception. Immediacy is derived' (157). As a result, the 'source' is an imaginary category:

> There is no present before [the supplement], it is not preceding by anything other than itself, that is to say by another supplement. The supplement is always the supplement of the supplement. One wishes to go back *from the supplement to the source*: one must recognize that there is *a supplement at the source*. (303)

Sebald's texts reflect a similar, though not identical, state of affairs. These considerations enter as a thematic concern in several passages of *Vertigo*: 'It was a severe disappointment, Beyle writes, when some years ago, looking through old papers, he came across an engraving entitled *Prospetto d'Ivrea* and was obliged to concede that his recollected picture of the town in the evening sun was nothing but a copy of that very engraving' (8). In this instance, memory (Beyle's 'recollected picture') is actually constituted as a reconstruction from an image, a representation – the purity of the unmediated vision is unattainable, and direct access to experience of the seen is in fact refracted through antecedent images. The text's narrator elsewhere understands this as a 'conjoining' of visual experience and previously seen reproductions or images: 'The low-lying cloud drifting in from the Alpine valleys and across that desolated country was conjoined in my mind's eye with a Tiepolo painting which I have often looked at for hours' (51). The site of this conjoining is the 'mind's eye', a composite that itself suggests the internalization of just such an assimilatory practice linking the sensory perception of the 'eye' with the thought processes and recollections of the 'mind'. These instances do not suggest, following Derrida, that there is no source that is not already supplementary; rather, they indicate that for Sebald (and for others on whom this book confers attention) literary inheritance and the moment of perception cannot be disentangled. Throughout *The Rings of Saturn*, the thorough-going permeation of perception of a landscape by its prior representations informs the very terms of description in which it enters the text, and so fundamentally underpins the understanding of landscape itself. In short, the landscape presented in the text is already bound up in narrative.

## *The Rings of Saturn*

W.G. Sebald, in *The Rings of Saturn*, returns numerous times to the works of the seventeenth-century physician and scholar Sir Thomas Browne (1605–1682), a figure he describes as having 'left a number of writings that defy all comparison' (9). Referring to individual texts by Browne, Sebald calls *Musæum Clausum* (1684) a 'register of marvels' (273), and emphasizes the genre-defying characteristics of Browne's output in finding *Urne-Burial* (1658) to be a 'part archaeological and part-mystical treatise' (11). Such a statement might also be applied to Sebald's own text. In addition to the stylistic influence of Browne on Sebald, *Urne-Burial* provides something of a thematic model for *The Rings of Saturn*: the earlier text comprises metaphysical and philosophical reflections prompted by the unearthing of Bronze Age burial urns at Walsingham in Norfolk. Browne's textual presence clearly goes beyond this thematic engagement: the earlier writer's work provides a paradigm for the structural operation of the poetics deployed in *The Rings of Saturn* itself. Indeed, the descriptive aesthetic and narrative style that seem to characterize the recent landscape writing with which this study is concerned are reflected in Sebald's reading of Browne. In another passage, Sebald describes Browne's writing style thus: '[He] wrote out the fullness of his erudition, deploying a vast repertoire of quotations and the names of authorities who had gone before, creating complex metaphors and analogies, and constructing labyrinthine sentences that sometimes extend over one or two pages' (19). The complexity of Sebald's own text is, at least in part, a result of the adoption of a register similar to that found in Browne. *The Rings of Saturn*, ostensibly an account of Sebald's walking tour of coastal Suffolk, employs this physical movement as the springboard for extended mental departures through his own 'vast repertoire', his archive of fascinations.

In this undertaking, Sebald blends personal memories with an array of textual sources – ranging from social and natural history, through cultural memory, to literary material – and assimilates the voices of these texts with his own writing in 'labyrinthine' constructions that, like Browne, employ 'complex metaphors and analogies'. If Sebald's labyrinthine texts are particularly idiosyncratic, they share with other contemporary writing on landscape an idea of place as constructed out of an amalgamation of extant resources and personal experience, as well as a stylistic tendency to recover the dynamic interpenetration of each by the other.

The emergence of strategies of writing and of reading employed by other writers as thematic concerns of *The Rings of Saturn* provides clues for deployment by the disoriented reader in grasping Sebald's own text. He is fascinated by Jorge Luis Borges's *Libro de los seres imaginarios* (1967), which he describes as a 'tale which deals with our attempts to invent secondary or tertiary worlds' (70). Sebald includes Borges's reflections on:

> the writing of a novel that would fly in the face of palpable facts and become entangled in contradictions in such a way that few readers – very few readers – would be able to grasp the hidden, horrific, yet at the same time quite meaningless point of the narrative. (70–71)

The passage is characteristic in drifting from a reportage mode into an appropriation of the first-person voice of the text paraphrased or sometimes even quoted from.[3] Similar examples abound in all of Sebald's texts and are often accompanied by convoluted sub-clauses and parentheses that emphasize the act of borrowing.

[3] To compare the original, see Jorge Luis Borges, *Labyrinths* 27. Sebald has described his penchant for silently 'borrowing an attractive image or a few expressions' as a 'mark [of] respect for those writers with whom I felt an affinity' (*A Place in the Country* 128).

This passage provides a model for reading Sebald's own 'entangled' and difficult text, which also has a 'hidden' and 'horrific' meaning, finding, as it does, that 'our history' is 'but a long account of calamities' (295). Sebald employs the category of the horrific as a mode of experience in sustained ways throughout the walks described. He feels a 'paralysing horror … when confronted with the traces of destruction, reaching far back into the past, that were evident even in that remote place [i.e. East Anglia]' (3). The relational conditions by which this meaning is arrived at, however, remain opaque: the fact that 'very few readers' may fully understand the text is a result of the extremely idiosyncratic assimilatory activity that it represents. Describing his colleague Janine Dakyns's reading of Flaubert, the narrator finds that her 'understanding of the nineteenth-century French novel … had about it a certain private quality, wholly free of intellectual vanity and was guided by a fascination for obscure detail rather than by the self-evident' (7). The disregard for the 'self-evident' here, much like the antipathy to 'palpable facts' in the Borges passage, alongside a 'fascination for obscure detail', is the guiding principle informing the selection of material in Sebald's texts. That is, with Browne and Borges as models, the narrator's perspective is situated not only in relation to the environment ostensibly described, but also in relation to the detailed artistic and cultural knowledge that he brings to that engagement. This fact is only further emphasized by the wide variety of languages and national traditions with which Sebald is evidently familiar.

Sebald also refers to Borges's text as a 'compendium' of 'imaginary beings' (22). This concept is instructive in relation to the organization of material in *The Rings of Saturn*. Its final chapter opens with an account of Thomas Browne's *Musæum Clausum*, which Sebald describes as:

> a catalogue of remarkable books … listing pictures, antiquities and sundry singular items that may have formed part of a collection put together by Browne but were more likely products of his imagination, the inventions of a treasure house that existed purely in his head and to which there is no access except through the letters on the page. (271)

Whilst Browne's text might have the appearance of factual cataloguing, the making of a compendium, it is actually an imaginative rendering of material in order to imbue it with meaning and to establish connections between discrete pieces of information. Retelling is therefore characterized as a creative act. Sebald's text is informed by this concept, constituting a kind of catalogue itself: Sebald constantly edges from description of a source towards appropriating its voice. His text incorporates numerous others as the narrator drops into the voice of Conrad, Chateaubriand, Michael Hamburger, and many others. The supplementary creative activity that Sebald himself engages in is the selection and assimilation of a huge variety of texts and narratives into his own. In *The Rings of Saturn*, the establishment of narrative is dependent on consignation and suturing material together. The curatorial activity of fixing the archive in a location is predicated on walking through an environment: the landscape so described functions as a catalogue of texts and images. But, as the prior invocation of the 'supplement' suggested, it is often the text (and the body of artistic representations that it orchestrates) that is anterior to the supposedly originary landscape itself.

It is upon the potentials for establishing narrative through the act of curation that the text dwells at length. Somerleyton Hall, near Lowestoft, is one of the first destinations of Sebald's walk. Open to the public, the Hall represents, for Sebald, a curated memorial space in which he experiences the co-existence of different times and places:

> There are indeed moments, as one passes through the rooms open to the public at Somerleyton, when one is not quite sure whether one is in a country house in Suffolk or some kind of no-man's-land, on the shores of the Arctic Ocean or in the heart of the dark continent. Nor can one readily say which decade or century it is, for many ages are super-imposed here and coexist. (36)

In textual form, Sebald's works exploit these same potentials for the making present of that which is absent. The eponymous character of *Austerlitz* uses the term 'stereometry' to describe the indiscernible laws by which different temporalities might interleave in particular spaces (*Austerlitz* 261). *The Rings of Saturn*, predicated on a walking tour of the East Anglian coast, incorporates all the associations that the visual experience of this journey brings to Sebald's mind. These associations rely on his own prior experience and reading and as a result our encounter with Sebald can only be as his is with Browne: 'there is no access except through the letters on the page' (271). The role played by reflection and curation in the creation of the text that Sebald's reader receives is further foregrounded by the emphasis placed on the almost Wordsworthian recollection-in-tranquillity of the journey across Suffolk being described. It is 'a year to the day after I began my tour' that the narrator 'was taken into hospital in Norwich [and] began in my thoughts to write these pages'. Furthermore, the actual composition of the text occurs after an additional temporal gap: it is 'more than a year after my discharge from hospital' that the narrator 'begin[s] to assemble my notes' (3, 5). The acknowledgement of multiple layers of textual production complicates the account provided in *The Rings of Saturn*, questions the reportage of unmediated experience, and lays stress upon its textuality and the processes informing the (re) construction of the landscape.

The central eight chapters of *The Rings of Saturn* are framed by opening and closing reflections that, in keeping with the focus on the act of curation, display a preoccupation with visual experience and the organization of images. It is of note here that the description of place in the text is largely undertaken by recourse to visual rather than other (sonic or tactile) stimuli. The photographs throughout the book affirm this priority. The privileging of the visual in the perception of space is a prevalent feature of Sebald's writing that others studied in chapters to follow go some way towards revising. The text opens with Sebald's narrator in a Norwich hospital and a description of the limited view that his bedside window affords:

> Indeed, all that could be seen of the world from my bed was the colourless patch of sky framed in the window. Several times during the day I felt the desire to

> assure myself of a reality I feared had vanished forever by looking out of that hospital window. (4)

The emphasis placed upon the physical limitation of the viewing position is here coupled with the concept of 'reality' to underscore the situated perspective from which the narrator sees. The narrator's initial placement within a hospital space is also significant in conjunction with the submerged suggestions of mental illness that permeate the entire text. By these means, an undertow of psychological disturbance is established as a means of further emphasizing the narrator's highly individuated perspective. This refrain is revived throughout, and most explicitly at the text's conclusion when it is suggested that scholars 'tended to suffer from melancholy and all the evils associated with it' (283). Even here at the novel's opening, this framed emptiness draws attention to the visual process and to the construction of seen things. An important link is established between seeing and believing. This economy is complemented only a few pages later with an elaboration of the problem of invisibility. Sebald looks from the same window: 'as the first light brightened the sky, I saw a vapour trail cross the segment framed by my window ... The aircraft at the tip of the trail was invisible as the passengers inside it.' For Sebald, this occurrence is an instance of the 'unfathomable mystery' of the 'invisibility and intangibility of that which moves us' (18). In these passages, the indeterminable relationship between the real and the imaginary, and the role played by visual experience and mental departure in this relationship, are established as concerns that permeate *The Rings of Saturn*, and the historiography at the core of the text, in significant ways.

The extent to which the observations that constitute a large proportion of Sebald's accounts of his walks emerge out of a combination of the imagined and the real is not disguised but emphasized throughout the text. Looking out to sea from his position at Gunhill in Southwold, the narrator recreates the historical event for which that view is famous:

> Everyone who had been out for an evening stroll was gone. I felt as if I were in a deserted theatre, and I should not have been surprised if a curtain had suddenly risen before me and on the proscenium I had beheld, say, the 28th of May 1672 – that memorable day when the Dutch fleet appeared offshore from out of the drifting mists, with the bright morning light behind it, and opened fire on the English ships in Sole Bay. (76)

In the terms of description, vision is linked to the theatre and the performativity that it suggests. The connotations attached to the theatre are those of a realm of illusion, but also of a space in which the imaginary is made highly material and tangible. It therefore provides a suitable metaphor by which to think about the dialogue that informs Sebald's approach to place. Landscape constitutes the liminal space in which the imaginary is made materially manifest. Also of importance is the fact that, for the narrator, the view becomes 'a *deserted* theatre' – the emphasis placed on the narrator's *solitary* contemplation illustrates the extent to which this is an idiosyncratic and personal reconstruction. Artistry mediating visual access is reaffirmed in Sebald's following observation that 'the pictorial representations of great naval engagements are without exception figments of the imagination' (76). These factors underlie the organization of Sebald's text: the walks that it reports upon provide the underpinning foundations for the construction of a narrative of mental departures; the common themes that these departures converge upon are drawn from characteristics read in the spaces that the text traverses.

Chief amongst the characteristics of East Anglia's coastal areas upon which Sebald's thoughts are predicated are its flatness and low population density that figure in his reading of the landscape as emptiness. The interplay between notions of emptiness and the process of re-filling that Sebald's bricolage of cultural memory engages in is crucial to the meanings established in his text. The pre-conceived objective of the walks, described in the novel's opening sentence, centralizes this concern: 'I set off to walk the county of Suffolk, in the hope of dispelling the emptiness that takes hold of me after completing a long stint of work.' Paradoxically, an antidote to internal emptiness is sought in a landscape that is characterized here as 'thinly populated countryside', or simply as a 'remote place' (3). This characterization on the part of the narrator is also affirmed in the historical sources that Sebald's text collates. For example, Mary Frances Fitzgerald is said to have found her Suffolk estate to be an 'impossibly remote place' (198). The perceived absences in the landscape offer the opportunity for the introduction of imaginary presences – it is upon these relational possibilities that the text is built. Walking between Walberswick and Dunwich, Sebald describes the landscape: 'The region is so empty and deserted that, if one were abandoned there, one could scarcely say whether one was on the North Sea coast or perhaps by the Caspian Sea or the Gulf of Lian-tung' (154). In this instance, the lack of features perceived in the East Anglian landscape introduces the possibility of relation to other times and spaces.

More commonly, however, initial emptiness is filled in with meaning now submerged or visually absent. In this way, Sebald's rendering of the spaces he describes relate them to overarching narrative tropes, chief among these is a historical narrative of decline and demise. This trajectory avowedly dominates the text, with the narrator constantly confronted by ever-present 'traces of destruction' from the outset onwards (3). As Berberich notes, 'Sebald's England is … a country defeated by the modern world, scarred by industries and centuries of abuse by her people' (177). John Banville goes further, finding that Sebald's writing aims to produce 'nothing less than a chronicle of the end of a civilization' (35–8). In this respect, *The Rings of Saturn* takes up fairly precisely where Sebald's first published work, the long narrative poem *After Nature* (1988, English trans. 2002), leaves off: its final sections are preoccupied with exploring 'the progress of history', depict a narrator 'plunged … into a quasi / sub-lunary state of deep / melancholia', and are concerned with traces of 'the previous century's ruins' in the East Anglian landscape (95–7). Overarching narratives and preoccupations are, of course, balanced against the geographical specificity suggested by 'traces' (located in place) throughout *The Rings of Saturn*. Sebald's description of the 'vast flatland' between Reedham and Lowestoft engages in this activity:

> Save for the odd solitary cottage there is nothing to be seen but the grass and the rippling reeds, one or two sunken willows, and some ruined conical brick buildings, like relics of an extinct civilization. These are all that remains of the countless wind pumps and windmills whose white sails revolved over the marshes of Halvergate and all along the coast until the decades following the First World War, one after the other, they were all shut down. It's hard to imagine now, I was once told by someone who could remember the turning sails in his childhood, that the white flecks of the windmills lit up the landscape just as a tiny highlight brings life to a painted eye. And when those bright little points faded away, the whole region, so to speak, faded with them. (30)

This is certainly not objective or neutral description. It conforms instead to the overriding themes of the text as a whole rather than necessarily being concerned to give an accurate description of a particular environment. Rather than offering documentary description, the passage gravitates towards typically Sebaldian fascinations (ruins, unpeopled scenes). The derelict windmills serve to illustrate a comparison between current emptiness and erased labour on the one hand, and the 'extinct civilization' of the past on the other. In the creation of this contrast, the actual visual experience of the present is supplemented by material drawn from reported reminiscence ('I was once told by someone who could remember') and from imaginary processes ('it's hard to imagine now'). The analogy with painterly technique ('as a tiny highlight brings life to a painted eye') emphasizes the degree to which creative devices are employed: the description is undertaken in terms that simultaneously deconstruct its own artifice and draw attention to the animation of the landscape. The process engaged in here, and on a larger scale throughout *The Rings of Saturn* is one whereby submerged meaning is recovered and articulated in a textual, narrated form of landscape.

In the narrative of decline that this passage suggests, the First World War, and elsewhere the Second World War in equal measure, provide significant points of departure and a constant referential framework. Sebald's own position, as a German living and working in, as well as writing about England is influential in this respect. The extent to which his personal position influences the recurring themes to which his depiction of the Suffolk countryside gravitates serves to place emphasis on the subjective viewing process and its situated perspective.[4] His interpretations of East Anglian locales draw on arcane local history but also frame

---

[4] The Second World War is a constant subtext in *The Rings of Saturn*. The submerged references to bombed-out wartime German cities are initially implicit in the description of Norwich as 'a field of rubble' at the text's opening, but later become explicitly related to the East Anglian subject matter in Sebald's conversation with the gardener at Somerleyton. The latter relays his memories of 'the bombing raids then being launched on Germany from the sixty-seven airfields that were established in East Anglia after 1940' (5, 38). War-torn cityscapes here recall the fascination they hold for Sebald as already established in *The Emigrants*: 'ever since I had visited Munich I had felt nothing to be so unambiguously linked to the word *city* as the presence of heaps of rubble, fire-scorched walls, and gaps of windows through which one could see the vacant air' (30).

these places in relation to diverse European literary and artistic traditions. Sebald has written elsewhere of his constant awareness of his foreign status throughout his time in England: 'Only a guest in England, I still hover between feelings of familiarity and dislocation there too' (*Campo Santo* 217). As Matthew Hart and Tania Lown-Hecht note, 'Sebald's extraterritorial poetics work in both directions: an empty coastline might be mapped onto a narrator's psychology, for example, but the narrator's sense of himself also becomes part of the landscape' (217). In the instance discussed here, as throughout the texts, his status as foreigner in exile impinges significantly on the reported experience of place.

The allusions to painterly technique are not the only means by which the landscape that the text describes is emphasized as one of representation as much as of sensory experience. The passage quoted above also provides a clear example of another formal aspect deployed throughout the text to this end. Whilst *The Rings of Saturn* is ostensibly an account of a walking tour, a 'pilgrimage' as the subtitle of the German edition suggests, it characteristically focuses expansively on static moments of vision that the walks lace together whilst passing over the actual moving between these specific sites. As such, the text presents a series of 'landscapes' in the technical perspectival sense derived from pictorial rather than the geographical usage – that is, '[a] view or prospect ... such as can be taken in at a glance from one point of view'. This introduction of the particular, fixed point of view, of perspective, emphasizes the subjective construction of the viewing process, and places it within a tradition of picturing or depicting landscape. The textual signification of this technique is also further underlined in the accompaniment of image alongside text in the page layout.

Photographs appear alongside text in this instance and throughout all Sebald's prose narratives. Their inclusion prompts questions regarding the relationship between and relative status of the various components presented on the page. Stephen Clingman notes that 'the photographs – on one level the form that emblematizes the very possibility of documentary – raise the question of verisimilitude in relation to veracity' (187). Simon Cooke goes further, finding that they offer 'contra-depictions' rather than 'corroborations': 'though the photographs lend an air of the documentary, this claim is as often dashed as not, offered as much in evidence, or as a subversion, of the narrator's viewpoint as much of any objective verification' (144). In his work on pictorial maps illustrating travel writing, Thacker provides a cue for thinking further about Sebald's photographs. As 'seemingly supplementary textual features to the central narrative of the travel book', a map '*adds* to the text and *substitutes* for the written text', and, as such, 'functions as a form of textual representation that alters how we read works of travel writing' ('Journeys without Maps' 11). Sebald's photographs, interrupting the text (often literally, mid-sentence) play a similar role. The image of the ruined windmill and flatland redoubles his textual description, underlining the uncertainties that already inhabit the latter – if the image serves documentary purposes alone, then it is implied that the text tells supplementary stories.

In addition, the photographs themselves seem to suggest their own artistry. The image of the flatland scene that accompanies the description above is carefully arranged so that the diagonal lines created by the watercourse and the overgrown walkway form a pair of organizing axes linking foreground and background. Additionally, the windmill on the right, as a picturesque ruin, makes reference to specific pictorial traditions. In short, the arrangement of objects in this photograph – the large sky, the waterway leading the eye from foreground to background, shrubs and trees, the building falling back into nature – seems to be contrived in a similar way to traditions that are as much concerned with artistic conventions as with objective depiction. At some compositional level, this image might be said to enter into dialogue with other famous Suffolk landscapes. In particular, John Constable's numerous paintings of mills and waterways – for example, *Dedham Lock and Mill* or *Flatford Mill* (both 1817) – bear comparison. Many other photographs depicting landscape or seascape in the book appear equally stylized. Given that one of the text's key themes is emptiness and solitariness, it is telling that so few of the pictures are populated. In the images, as in the text, dialogues and intertextualities are established specifically and generically, with particular texts or images and to pervasive traditions and structures. Often, both species of reference are occurring in tandem, further developing the complex web of interrelations in which place exists and is perceived. The status of Sebald's landscape writing – and the interplay of representation and experience therein – are modified significantly by these factors. This is not to suggest that sensory experience in the present is diminished, but that it occurs in and is inevitably conditioned by multiple cultural contexts.

The photograph of the flatland scene, like many others, confronts the reader with visual evidence of the temporary fixing of landscape in a particular representational perspective. However, the fact that the text and images record static moments of vision does not preclude movement and mobility; rather, it suggests the altering rhythms that pausing to look at a view incorporates into any walk. David Darby argues that '[m]ovement through landscape is essential to the process of memory enacted in Sebald's writing' (265). In this context, an important distinction emerges: the text's views are pauses – moments of stillness – rather than extended stasis. *The Rings of Saturn* thus explores the play between stability and dissolution. Its situated yet mobile perspective lies somewhere between fixing and holding open: the fixing of the landscape occurs in an idiosyncratic manner that draws attention to itself, so that the process of visualizing is construed as part of individuated experience of environment. In short, the text foregrounds the mutual interdependence of representation and sensory experience of place. Sebald's work can thus be read in line with Crouch's comments on the relationship between representations and lived experience. Whilst the former are 'familiarly considered as objects, objects of their own completeness that have reduced landscape to available contemplation', they are also 'momentary expressions of journeys' wherein 'combinations of influences and affects erupt'. 'Landscape', for Crouch, and undoubtedly for Sebald as well, 'emerges in moments of diverse temporality and different intensities' (11–12).

These concerns also explicitly enter the thematic content of the text on the several occasions on which the narrator visits galleries and views particular paintings (often, but by no means always, landscapes). 'If we stand today before the large canvas of Rembrandt's *The Anatomy Lesson* in the Mauritshuis', Sebald writes, 'we are standing precisely where those who were present at the dissection in the Waaggebouw stood, and we believe we see what they saw.' However, the mimicry of perspective and patterned ways of seeing that this artistic representation provides are found to be problematic: 'the much admired verisimilitude of Rembrandt's picture proves on closer examination to be more apparent than real' (13, 16). The distorting, that is, mediating, role played by art is then emphasized in the reading of the painting that Sebald goes on to provide. Rembrandt's sympathy for the victim of the dissection rather than his alliance to his patrons, the Guild of Surgeons, is apparently indicated in his deliberately disproportionate and anatomically incorrect depiction of the cadaver's hand. This fact 'turns this otherwise true-to-life painting … into a crass misrepresentation at the exact centre point of its meaning' (16). Sebald's disquisition insinuates difference at the heart of the supposed mimetic representational act. Though the interpretation that he provides is not necessarily convincing, the point is again one of distinguishing between verisimilitude and veracity.

The description of Ruisdael's *View of Haarlem with Bleaching Fields* is equally loaded:

> The flatland stretching out towards Haarlem is seen from above, from a vantage point generally identified as the dunes, though the sense of a bird's-eye view is so strong that the dunes would have to be veritable hills or even modest mountains. The truth is of course that Ruisdael did not take up a position on the dunes to paint; his vantage point was an imaginary position some distance above the earth. (83)

'Flatland', the same term used to describe Sebald's East Anglian landscapes, is employed here and thus serves to emphasize the necessity of interrogating the perspectives from which those views are achieved. Indeed, the selection of a Ruisdael painting for making this observation refers back to Suffolk – Ruisdael was, as Kenneth Clark notes, 'to become the model for the whole East Anglian school of landscape painting' (46). This is only one instance of an extensive set of correspondences between Dutch landscape and history and the English region through which Sebald walks. (As the text makes clear, the two face one another across the North Sea, and have come into mercantile and military contact at numerous historical moments.) The elevated, panoptic, fictional perspective is often described in negative terms. The selection of the comparably flat Suffolk coastline for Sebald's walk emphasizes a distinction to be made here. Recalling a visit to the site of the battlefield at Waterloo, the narrator details the raised platform providing panoramic views: 'This then, I thought, as I looked round about me, is the representation of history. It requires a falsification of perspective.' The result of this is that we 'see everything at once, and still we do not know how it [the battle] was' (125). In these instances, whether in Rembrandt's insertion of subversive meaning or in the formation of all-seeing perspective by Ruisdael and at Waterloo, the situation of perspective is emphasized as crucially important in the formation of meaning, and, particularly in the latter example, of a historical narrative.

At many points in *The Rings of Saturn*, the deliberate emptying of the described view can be seen to provoke the narrative that Sebald reads there. After eating a meal in a Lowestoft guesthouse, Sebald describes the view from the window:

> When she [the landlady] had flitted away once more I rose and crossed to the semi-circular bay window. Outside was the beach, somewhere between the darkness and the light, and nothing was moving, neither in the air nor on the land nor on the water. Even the white waves rolling in to the sands seemed to me to be motionless. (43)

The view is framed by the prior description of the window and its stillness is emphasized in contradistinction to the landlady's 'flitting'. The construction of the image itself is suggested in the enumeration of its component features, with the list effect also serving to reduce the pace of the description and foreground emptiness. Following this passage, the text jumps to a description of Lowestoft the following morning. The town, with its 'dozens of decommissioned and unemployed trawlers', and the apparently spontaneous yet symbolic appearance of 'a black hearse', is characterized as one 'on which the marks of an insidious decay were everywhere apparent, a town which in its heyday had been not only one of the foremost fishing ports in the United Kingdom but also a seaside resort lauded even abroad as "most salubrious"' (44–5). The 'dozens of decommissioned and unemployed trawlers' here, and lack of presence or practice in the landscape descriptions given on previous pages, are comparable to the emptying of the signs of labour from the landscape that Raymond Williams charts at length in *The Country and the City*. This factor contributes significantly to Sebald's narrative of decline and punctuates the text at regular intervals: upon the narrator's arrival in Lowestoft, '[n]ot a living soul was about in the long streets I went through'; viewing the countryside around Amsterdam from a plane window, '[n]owhere … was a single human being to be seen'; crossing the river Blyth, '[t]oday there is next to no traffic on the river' (41, 91, 137).

The effect of the Lowestoft sequence is that the aura of emptiness created in the initial description is filled with an unearthed narrative of decline in the following pages. The general tendency of the text as a whole, built up in instances such as this, is to read a narrative of human history in the renderings of place and space that Sebald engages in. As a further result, Sebald makes a repeated movement from the particular to the general. Thus his observations of Lowestoft – 'it seemed incomprehensible to me that in such a relatively short period of time the place could have become so run down' – lead inexorably in the following sentence to macro-narrative history – 'I was aware that this decline had been irreversible ever since the economic crises and depressions of the Thirties' (41). The relay from the particular to the general, and therefore the relation that they bear to each other, is via a complex interaction of space and the individual's reading of that space.

In the service of Sebald's selective narrative strategies and the sense of decay that they feed into, the varying pace of the text, along with the walk that it reports on, selects certain sites (and sights) for visitation and prolonged analysis. These present a history and/or topography suited to this mode of reading, and the frequency with which Sebald's itinerary encounters them might suggest that the apparent randomness with which the trips are presented in the finished text is less spontaneous than it seems. This factor goes some way towards disproving Will Self's reading of the text. He argues for the journey as the genesis of the associated narratives: 'These were picaresques, pure and simple: the journeys were the foundation of all the tales, the embellishments – personal, anecdotal, historical – came along the way, called forth by incidents on the road' (3). Finding the level of coincidence and the imposition of an all-pervasive grand narrative too obviously contrived, I fundamentally invert Self's reading and argue for a textual genesis of the journeys: Sebald's itinerary is carefully planned to take in historically resonant locations. The text, then, is coloured by a specific historiography of decline embedded in space, place, and landscape.

Contra to Self's notion of picaresque, the solemn journey to resonant sites recorded in the text might more accurately be described as a kind of pilgrimage, as the subtitle of the German edition ('eine englische Wallfahrt', an English pilgrimage) makes clear (*Die Ringe des Saturn*). J. Stopford has suggested that across a diverse range of ritual practices encompassed by the term 'pilgrimage', all 'involve reference to the past, recalling previous lives and events which are believed to be applicable in the present' (x). This is certainly true of Sebald's pilgrimage – the landscape that it traverses is saturated with a history that feeds into the narrative of demise of the text's present. In this context, the inquiry into the whereabouts of Sir Thomas Browne's skull in an early section of *The Rings of Saturn* might be considered a kind of search for relics (9–11). Browne's is certainly one of the lives that Sebald looks to connect with. The text's antiquarian interests are buttressed by the adoption of a mode of walking associated primarily with a much earlier, medieval period. The idea of pilgrimage here is repeated in other writers' texts read in subsequent chapters of this book. Indeed, the notion of sacred places has become something of a leitmotif for contemporary landscape writing. In the context of this study's wider interests, Alexandra Walsham's understanding of pilgrimage as 'the external expression of an inner quest for transcendental meaning' suggests some of the ways in which cerebral and corporal experiences of place interact (49). Journeying to a specific place for its associations constitutes a practice of environment informed by inherited ideas and representations.

If Sebald's walk is a pilgrimage, its shrines are more often than not ruins. Pilgrimage here, then, is a secular form of ritual in its conferral of attention on sites of human significance (even if these are sometimes churches), and is melancholic rather than celebratory in seeking out remnants of entropy. Dunwich is given extended treatment for these reasons:

> The Dunwich of the present day is what remains of a town that was one of the most important ports in Europe in the Middle Ages … All of it has gone under, quite literally, and is now below the sea, beneath alluvial sand and gravel, over an area of three square miles. The parish churches … one after the other, toppled

> down the steadily receding cliff-face and sank in the depths, along with the earth and stone of which the town had been built. (155)

The present settlement is read as the 'remains' after demise from 'the high point of its evolution in the thirteenth century' (157). The process of coastal erosion allows Sebald to employ a depth model to demonstrate the town's destruction. Darby has noted the role played by the literal disintegration of the Suffolk coastline in *The Rings of Saturn*. Sebald's journey, he argues, is 'a walk literally on the edge of an abyss, the eroding terrain ever crumbling away beneath the feet of the walker, dissolved by the elements' (270). Darby goes on to argue that the landscape of the text is in a state of flux – its only afterlife surviving in the memories attached to it. The inclusion of description in a geological lexis ('alluvial sand and gravel'), and of the area's dimensions ('three square miles') in this passage suggest scientific disciplines in his mode of recovering the lost Dunwich: Sebald's writing takes on the characteristics of a textual reconstruction comparable to archaeology. Erosion, a key natural process in the formation and modification of the East Anglian landscape, becomes, for Sebald, a key motif to which a supplementary meaning is superadded; the text shifts from here towards describing the tendencies of a human history:

> Over the centuries that followed, catastrophic incursions of the sea into the land of this kind happened time and again, and, even during the long years of apparent calm, coastal erosion continued to take its natural course. Little by little the people of Dunwich accepted the inevitability of the process. They abandoned their hopeless struggle, turned their backs on the sea, and, whenever their dwindling means allowed it, built to the westward in a protracted flight that went on for generations; the slowly dying town thus followed – by reflex, one might say – one of the fundamental patterns of human behaviour. (158–9)

The gradual movement of this description is away from its opening focus on natural processes and towards human ones. The boundary between the two is thus blurred and the transition from one to the other smoothed. In this example, Sebald's tendency to treat 'nature' and 'human nature' as synonyms is demonstrated. Thus, looking 'from the cliff-top across the sea to where the town must once have been, you can sense the immense power of emptiness' (159). The compelling power of emptiness lies, for Sebald, in submerged history and the potential this creates for reading to act as re-filling.

The same patterns are repeated when the narrator reaches Orfordness. A description of the natural processes forming the features of the place are present – 'Stone by stone, over a period of millennia, it had shifted down from the north across the mouth of the River Alde, in such a way that the tidal lower reaches, known as the Ore, run for some twelve miles just inside the present coastline before flowing into the sea' – but are just as soon passed over (233). Sebald's focus dwells not on these features but on the psychological effects of the abandoned military infrastructure to be found there (another set of ruins). He finds that the local fishermen 'couldn't stand the god-forsaken loneliness of that outpost in the middle of nowhere, and in some cases even became emotionally disturbed for some time' (34). The reading presented in *The Rings of Saturn* follows the trajectories already well established by this point in the text – 'With each step that I took, the emptiness within and the emptiness without grew ever greater and the silence more profound' – and Sebald's analysis of the visual landscape finds that 'ahead lay nothing but destruction' (234–5). The area offers the grounds for an extension of the narrative of decline into an imagined future apocalypse: 'But the closer I came to these ruins, the more any notion of a mysterious isle of the dead receded, and the more I imagined myself amidst the remains of our civilization after its extinction in some future catastrophe' (237).

The meaning that is read, or perhaps written, in the environment with which Sebald's text deals is primarily associated with human history – with the people who have dwelt at or visited the place, and the historical events for which it is remembered – and less with the processes of the environment itself. The two are not unrelated, but the relationship is one of inverse fortunes: 'Our spread over the earth was fuelled by reducing the higher species of vegetation to charcoal, by incessantly burning whatever would burn' (170). Exploiting this fact, on the occasions when the fate of the natural landscape enters the text without explicit human meanings attached, it does so as a further marker of the narrative of human decline with which Sebald is primarily concerned. Thus, Sebald's report of the grounds surrounding the manor house at Henstead is included not to celebrate the success of plant life, but to emphasize the decline of the manor's inhabitant: 'the whole estate, with its gardens and park, became overgrown and neglected, while scrub and undergrowth encroached on the fallow fields' (62). The same pattern occurs again, with the attachment of the same meanings, at Somerleyton – '[t]he grounds, in contrast to the waning splendour of the house, were now at their evolutionary peak, a century after the heyday of Somerleyton' – and in the description of the land between Woodbridge and Orford – '[w]hen the shepherds and their flocks disappeared in the early nineteenth century, heather and scrub began to spread' (37, 222). In this last instance, Sebald is playing off pastoral conventions as much as historical fact.

In the face of the insurmountable and all-pervading narrative of decline and demise that it presents, what *The Rings of Saturn* actually attempts to enact is a textual recovery of that which is passing. For Sebald, it is in the writing and reading of texts that the interplay of emptiness and re-filling finds its resolution. The problems arising where there is a lack of written records are described in relation to the Second World War: 'No one at the time seemed to have written about their experiences or afterwards recorded their memories. Even if you asked people directly, it was as if everything had been erased from their minds' (39). The very process of erasure that the landscape suggests is cast as a result of the lack of textual records.

Elsewhere, however, the presence of texts offers redress to the absences in the landscape. Whilst in the Sailor's Reading Room at Southwold (a significant location in this respect), Sebald looks through 'the log of the Southwold, a patrol ship that was anchored off the pier from autumn of 1914':

> On the large landscape-format pages, a fresh one for each new date, there are occasional entries surrounded by a good deal of empty space, reading, for instance, Maurice Farman Bi-plane N'ward Inland or White Steam-yacht Flying White Ensign Cruising on Horizon to S. Every time I decipher one of these entries I am astounded that a trail that has long since vanished from the air or the water remains visible here on the paper. (93)

The 'landscape-format pages' offer a double to the actual landscape surrounding the library, and the text contained in the log is literally adjacent to the 'empty space' surrounding it on the page. The erasure enacted in the landscape is supplemented by textual sources. Here this process is explicit, but it also implicitly underpins the entirety of *The Rings of Saturn* in the body of knowledge related to East Anglia that Sebald has clearly built from textual sources prior to commencing the walks themselves. In this way, to again re-invoke Derridean terminology, the supplementary activity (textual research) is in fact the genesis of the engagement with the actual landscape. In short, the supplement is at the source (Derrida 303).[5] *The Rings of Saturn* thus focuses upon what is, for Sebald, 'the mysterious survival of the written word' (93). The role played by texts in the movement from emptiness to fullness is also related to another recurring pairing pervading Sebald's work – that of chaos and order. Describing Janine Dakyns's office, Sebald observes the 'ever increasing masses of paper' that fill it. Significantly, her desk is described as a 'virtual paper landscape'. However, the superficial impression of disorganization is revised: 'the apparent chaos surrounding her represented in reality a perfect kind of order, or an order which at least tended towards perfection' (8–9). Just as occurs in Sebald's landscape descriptions, surface chaos is actually underlain by order. Here the significance is the implication of papers and texts in this process. Sebald's own text might be read as an imposition of this transition on the East Anglian spaces that he traverses – his text imposes signification and order, recovering their meanings, and (re)presenting the landscape in textual form as an archive of its history.

As such, *The Rings of Saturn* constitutes a further development of these concerns as they exist in his earlier works. In an essay on Rousseau, Sebald comments on the writer's need to instil order as a kind of curse: 'Thus this apparently innocent occupation – the deliberate resolve no longer to think and merely to look at nature – becomes, for the writer plagued by the need to think and work, a demanding rationalistic project involving the compilation of lists, indices and catalogues' (*A Place in the Country* 55). *Vertigo* displays the same fascination with the potentials for imposing order on an unruly world in a textual realm offered by the creation of archives. Closely following the account of a disturbing encounter with a real environment, that text's narrator describes the contrasting calm to be found in texts: 'August the 2nd was a peaceful day. I sat

---

[5] A sizeable sub-section of Sebald criticism has grown up searching out and detailing the myriad textual sources at the genesis of the narratives. For example, see Bianca Theisen, 'Prose of the World'.

at a table near the open terrace door, my papers and notes spread out around me, drawing connections between events that lay far apart but which seemed to me to be of the same order' (94). Here, the imposition of order brings pleasure. It takes place in a textual realm of 'papers and notes' and is reliant upon an idiosyncratic mode of relation: the narrator connects those things 'which *seemed to me* to be of the same order'. This is not a statement of fact but an expression of appearance. However, the failure of this imposition of textual order follows quickly: 'The writing was becoming increasingly difficult, and soon it all seemed to be the most meaningless, empty, dishonest scrawl' (95). *Austerlitz*'s long narrative concludes with a similar frustration at the inability of archives to offer full compensation; instead he suggests that 'the dissolution, in line with the inexorable spread of processed data, of our capacity to remember' is occurring (398). The fragility of the archive's order is thus implied. The narrator of *Vertigo* admires a similar textual order in an Italian grammar primer previously belonging to his uncle: on its pages 'everything seemed arranged in the best of all possible ways, quite as though the world was made up purely of letters and words and as if, through this act of transformation, even the greatest of horrors were safely banished, as if to each dark side there were a redeeming counterpart, to every evil its good, to every pain its pleasure, and to every lie a measure of truth' (105). Text, in this conception, becomes a shelter from the chaos of reality: 'Once I am at leisure, said Salvatore, I take refuge in prose as one might in a boat' (*Vertigo* 128). The relationship between text and reality constitutes the core of meaning emphasized at length in *The Rings of Saturn*.

## The Spatial Supplement

Mapping out the thematic engagements of Sebald's texts, J.J. Long and Anne Whitehead have found him to be 'intensely concerned with questions of representation, with the forms in and through which individual and collective memory, historical knowledge and the cultural heritage can be communicated from one person to another and passed down through the generations' (4). This fact, they note, has implications for the formal characteristics of Sebald's style of writing: by 'drawing attention – by explicit or implicit means – to his own representational strategies, Sebald thematizes the question of transmission of historical knowledge and personal and collective memory' (5). Similarly, John Marks has described Sebald's project as an 'attempt to construct a literary form that confronts the problems of representation in the wake of much recent European history' (89). These 'representational strategies', as the above analyses describe, are largely predicated upon the establishment of a particular and idiosyncratic 'situated perspective'. This, in turn, has involved Sebald's deployment of landscape description to create opportunities to comment on the shortcomings of traditional historiography and suggest alternative historiographical perspectives.

A final invocation of the Derridean concept of supplementarity might be employed to describe the ultimate function of spatial perspectives in Sebald's historiographical critique. The relation of 'space' (and its cognate associations signified in 'spatial') to 'history' (and the 'temporal'), as they figure in Sebald's texts, is that of the former as supplement to the latter. In this relation, both of the Derridean functions of the supplement are fulfilled. First, the spatial is simply added to (supplements) the historical, and it is out of a complex of the two that events and narratives occur. Second, the spatial constitutes an interrogative category that offers an alternative positioning in relation to historiography – a substitute perspective upon its commonplaces that comes to redefine its core. Space's supplementary (that is, substitutionary) role is not simply one of replacement but of perspectival displacement or refraction. Space, and the 'spatial turn' in arts and humanities disciplines, offer a *new way of seeing* history. What the spatial turn is not, in any simple way, is an escape from (a turn *away* from) historiography; rather, it reflects upon (turns *in* on) historiography.

John Wylie's reading of Sebald's texts employs a Derridean model of 'spectrality'. 'The ghostly, or the spectral', Wylie notes, 'is, always, *revenant* … The spectral is thus the very conjuration and unsettling of presence, place, the present, and the past.' Whilst this has significant implications for the conception of 'place', Wylie himself notes its implications for time: 'the spectral above all confounds settled orders of past and present' ('Spectral landscapes' 171–2). The crucial point is that the spatial and the temporal interact. My own analysis has aimed to demonstrate that sensory perception of landscape is, in *The Rings of Saturn*, haunted by the constantly revenant history of representations of those same (and other) landscapes. Just as literary critical research profits from an engagement with cultural geography, the latter might also profit from the focus on narrative technique offered by the former. In this way, the event and processes of inhabitation might be understood in *plotted* contexts. Grounded in this composite critical paradigm, this case study of Sebald's texts has implications for the ways in which landscape might be conceived. It demonstrates how landscapes might be narrated without simply being reduced to texts. It also offers a challenge to the treatment of representation as a primarily pictorial tradition – instead it focuses on the interpenetration of the visual and textual, and emphasizes a literary tradition that needs to be accounted for. In short, it focuses upon the ways in which literary landscapes are always caught up in narrative.

# Chapter 2
# Mapping: Ciaran Carson

This chapter opens with a comparison of two statements to sound several of its key notes. Rebecca Solnit: 'Walking the streets is what links up reading the map with living one's life, the personal microcosm with the public macrocosm; it makes sense of the maze all around' (176). Ciaran Carson: 'Perhaps the way out of the labyrinth is to get deeper into it, more fully to explore its ramifications' (qtd. in Kennedy-Andrews, 'For all I know' 25). In the first, Solnit remarks on the way in which a particular practice of space – walking – constitutes a means by which to bring together and establish dialogues between the representational (mapped) and experiential (lived) registers of a place. In this way, the maze is 'made sense of': what was confusing becomes clear. Walking offers resolution. The second statement, given by Carson in an interview, might be read in a similar vein. Here, exploring – an activity that, for Carson, would most often take place on foot – plays a comparable role, offering reconciliation if not resolution. However, Carson's terminology and phrasing is characteristically ambiguous, even paradoxical: to find the way out of the labyrinth is also to move deeper into it. If, for Solnit, the maze-like qualities of urban space are to be overcome, then for Carson, they are also to be embraced. Carson's purpose, I hope to show, has often been to recognize the complicated interaction of the various perspectives on place that are inscribed in variant modes of experiencing and representing urban landscape.

Critical responses to Carson's work have not focused exclusively on these issues. Rather, literary geography has been only one of a number of critical projects, though it has come to the fore in this field more recently. Northern Irish politics and history has also played a large part in conditioning reception to date. David Wheatley has noted that 'the Troubles have provided an inevitable and tragic backdrop, warping their social deformations into the fabric of Northern Irish writing' ('That blank mouth' 4). If the sectarian conflict has indeed been an 'inevitable' context in which this poetry has been written and read, it need not be the only one. For a study invested in analyzing representations of place, the territorial character of this dispute and the sectarian geographies that it perpetuates allow for a dovetailing of concerns. However, there are also drawbacks to this particular politicized critical approach. Fran Brearton has noted that '[t]o read poetry according to the imperatives of time and place (the Troubles, Northern Ireland) is also too often to miss the broader poetic context in which that work should properly find its "place"' (95). Readings of Carson's work exclusively in relation to a Northern Irish context run the risk of neglecting other possible ways of locating its meanings.

It is perhaps partly in response to these concerns that Carson scholarship has increasingly tended to read him primarily as a poet of place, of the city, and specifically of Belfast, rather than as a poet of the Troubles (though, of course, the two arenas are intimately related). *Ciaran Carson: Critical Essays* (2009), edited by Kennedy-Andrews, makes clear the extent to which Carson's texts speak to literary geographical concerns with numerous essays offering titular focus on space, place, and landscape.[1] Neal Alexander's monograph *Ciaran Carson: Space, Place, Writing* (2010) emphasizes the same set of concerns. Furthermore, wide-ranging surveys of the relationship between poetry and place have engaged with Carson: Peter Barry, in *Contemporary British poetry and the city* (2000), emphasizes the extent to which his poetry is 'relentlessly loco-specific' (*Contemporary British poetry and the city* 229); and Eric Falci selects Carson as one of three writers paradigmatic of post-war British and Irish poetry's response to landscape. A number of critics have also elaborated critical methodologies for apprehending Carson's texts that draw on paradigms from cultural geography. Julia Obert, for instance, finds Carson's poetry to be 'insistently phenomenological' resulting in an 'acoustemological poetics' (1082, 1084). In a similar vein, Jonathan Stainer considers Carson's 'subject-enriched and perspective-enriched' geographical mode to be a pluralist political tactic (373). My own literary geographical approach does not undertake exhaustively contextualized reading or offer extensive comparisons to other Northern Irish poets of the same milieu (though these links are undoubtedly fruitful). However, the Belfast that Carson depicts, along with the mode of that depiction, is permeated by the Troubles in thoroughgoing ways. As a result, this chapter aims to draw on historicized accounts of Carson's writing and to elucidate the ways in which his poetry of place and his spatial poetics have developed in response to a specific set of circumstances.

Building on ideas of 'situated perspective' established in the previous chapter, I aim here to show that Carson's poetry emphasizes the uncertainties of any attempt to register Belfast's places. Sarah Broom describes the 'the intense external pressure to comment on "the situation"' and to take up a (political) 'position' that has been felt by Northern Irish poets (142); and Shane Alcobia-Murphy has detailed the tendency to 'eschew definitive statements' that has informed their work as a response (32). Instead, Carson and others have, in varying forms and by various means, employed a poetics of obliquity in response to the binary logic of the conflict. I argue that Carson's poetry balances a tendency towards indeterminacy against an equal fascination with the possibility for a recuperated Belfast before, outside, or beyond the established narratives of the Troubles. Amongst his contemporaries, Carson has offered the most sustained exploitation

---

1 The volume includes: John Goodby's '"Walking in the city": space, narrative, and surveillance in *The Irish for No* and *Belfast Confetti*', Eamonn Hughes's '"The mouth of the poem": Carson and place', Jerzy Jarniewicz's 'Alphabets and labyrinths in Ciaran Carson's *Fishing for Amber*', and Kennedy-Andrews's own essay, 'Carson, Heaney, and the art of getting lost'.

of spatial dynamics to question the idea of a political 'position' and to undercut the pressure to take up a 'fixed' (that is, static) stance.

Carson's oeuvre offers a sustained focus on the cityscape of Belfast from *The Irish for No* (1987) and *Belfast Confetti* (1989) to more recent figurations in poetry and prose such as *The Star Factory* (1997), *Breaking News* (2003), and *For All We Know* (2008). If, on occasion, he has criticized other writers for failure in 'seeing what's before your eyes', these texts have engaged extensively and specifically with the materiality of Belfast, and developed a poetics interrogating ways of seeing ('Escaped from the massacre?' 186). He challenges the imperatives of taking up a position by emphasizing the problems of representation implicit in the contemporary sectarian geography of Belfast and the cartographic thinking that underpins it. In contrast to the landscape of checkpoints and the static identities that they enforce upon Belfast's inhabitants, Carson interrogates mapping processes to reveal the multitude of perspectives and perception present in a polyvalent city. The focus on 'picturing' in Chapter 1 carries over into a consideration of 'mapping' here: both terms might be said to resonate processes of perceiving and of conceiving. This chapter thus concludes with a consideration of Carson's employment of a camera metaphor to describe the processes of recording and redrafting that inhere in his representations of Belfast.

Carson's poetry operates within a Lefebvrean notion of the production of space as ongoing, open to interrogation and intervention. It is in light of this interest in the instabilities of place and of perspective that Carson has engaged extensively with ideas of mapping and cartography. These same concepts have also assumed a crucial role in recent interdisciplinary critical work. Indeed, the currency of 'mapping' as a term has expanded vastly in new research. Daniels et al. have noted that the term 'has moved beyond the practice of cartography to a broader, metaphorical sense of interpreting and creating images and texts and making sense of a fast modernizing or post-modernizing of the world' (xxx). This shift is paradigmatic of the dialogues that current research is establishing between practices and representation. The prevalence of ideas of mapping has run parallel – at least in its applications in the humanities – with an interrogation of the supposedly objective discipline of cartography that employs scientific procedures and equipment to produce printed maps.

Alexander suggests that the ambivalence characterizing current understandings of mapping emanates from its polarized set of references. On the one hand, the term is equated with 'the apparatuses of the state and of social control, as a sort of graphic inscription of panoptic authority that is imposed on the territory'; whilst on the other hand, 'the map is conceived in terms of performativity, as a tool … capable of constant modification and multiple uses rather than imposing a monolithic order' (*Ciaran Carson* 57–8). In the terms of this monograph, then, Carson's fascination with mapping can be seen as an intervention in the relationship between experience and the particularly freighted form of representation that the term simultaneously describes. Whilst a map purports to provide an accurate description of what one might find on the ground, it does so by highly symbolic means and through an evolved system of codification. Maps are a form of representation susceptible to the same pitfalls as all other representational mediums: they are unstable, carry particular ideological impetus, and are to be interrogated rather than accepted. The critical tendency to conceive of maps in this way emerges broadly contemporaneously – around this time, David Harvey was suggesting that cartography is 'far from ideologically neutral' (228; see also Harley). Because of this mediation, Carson, like others, is suspicious of the map's ability to represent places and landscapes as they are experienced and practised on the ground. However, this does not mean that he is always censorious in his dealings with maps. Rather, their inconsistencies often figure in his poetry and prose as productive and pleasurable. Indeed, the coded and symbolic characteristics of maps are likened by Carson to literary practices of representation and especially the poem. In interview, he has suggested that '[f]or a map to work, it has to use shorthand, or symbols, or metaphor, and in this it resembles poetry' (qtd. in Ormsby 5). Carson's texts emphasize similarities: for example, chapters of his prose work *The Star Factory* (1997) are titled with Belfast street and building names, associating the text with the guide and the map. In short, Carson works to unpack and recalibrate the diverse associations of mapping as it has recently been understood, and to draw together practices of space and the representations that emanate from them.

At the heart of Carson's poetry is an elaboration of, and resistance to, the disciplinary imperatives of a primarily cartographic apprehension of space. His representations of military and paramilitary uses of mapping and the discipline of cartography permeate his poetry and prose. However, like almost all concepts in Carson's writing, the map is not limited to a single context. His writing develops a series of versions of mapping processes and plays one off against another. 'Hodology' or path-finding, mental mappings, street-level perspectives, and maps of memory are shown to intersect with (and intervene in) the dominant, disciplinary procedures associated with cartography. Carson's poetry works to reveal the complex interaction of these different mappings and, ultimately, to hold authoritative versions of the process open to revisionary impetus from alternative positions. Here, no single version prevails. If the authority of the map is discredited, this chapter argues that the labyrinth provides a more dynamic alternative. This motif, both as an alternate conception of the city's fabric and as a figure of narrative structure, constantly undermines positions of certainty and the all-encompassing, definitive perspectives that others in Northern Ireland claim. To this end, Carson develops a particular set of labyrinthine, woven aesthetics of *the space of the text itself*. His writing is labyrinthine in rhythm and structure as well as in frequent allusions to Irish, English, and other European language traditions. Like Sebald, Carson looks beyond nationally defined cultural boundaries. In elucidating these points, I aim to account for the traffic between material and metaphorical space; that is, in Thacker's terms for literary geography, to account for the 'interaction between spatial forms and social space in the literary text' (*Moving Through Modernity* 4).

## The Map

*The Star Factory*, Ciaran Carson's prose narrative that moves between unconventional autobiography and account of Belfast's recent history, addresses at length the issues of replication relevant to the city's representation in cartographic form. In an early passage, it does so through the metaphor of model aeroplanes: 'Sometimes the city is an exploded diagram of itself, along the lines of a vastly complicated interactive model aircraft kit whose components are connected by sprued plastic latitudes and longitudes' (15). The correlation between the actual city and its diagrammatic representation, tentatively described as 'sometimes' viable and subject to 'vast complication', is made more explicitly problematic in the sentence following on from this: 'At the same time it mutates like a virus, its programme undergoing daily shifts of emphasis and detail' (15). Mutations occurring daily in the material city severely curtail the temporal limits within which the map is accurate. Carson goes on to express admiration for 'true believers in mimesis', but finds that '[t]hey are in a parallel reality' (18). Conceptual, geometric apportioning of space (particularly resonant with respect to the attempted partitioning of Belfast's neighbourhoods during the Troubles and since) contrasts everyday experience of the city. Carson's texts typically respond to the question of the relation between a 'parallel' world and actual 'reality' implicitly being posed here. Historical attempts to fix the map of Northern Ireland, and their failure, are recorded. 'The Brain of Edward Carson' envisions 'The map of Ulster opened up, hexagonal and intricate, tectonic: / Its shifting plates were clunked and welded into place', but goes on, at the poem's conclusion, to chart how 'disintegration intervened' (*Ballad of HMS Belfast* 102). In these lines, an extended geological metaphor invokes and simultaneously undermines stasis. Elsewhere, Carson's writing offers a more sustained focus on the contemporary city and formulates specific forms of 'disintegration' and 'intervention' (versions of 'opening up') in such mapping processes.

'Turn Again' appearing at the head of *Belfast Confetti*, is set apart from the other poems of this collection. As such, this poem constitutes a kind of epigraph indicating the thematic concerns of the sequence. It is explicit and extensive in its engagement with mappings. The poem's opening lines find a stable cartography to be an unattainable ideal incommensurate with the reality of place as a site of change. Lines one and two detail imagined cartographies of the future aspired to but never achieved, and of the past now altered: 'There is a map of the city which shows the bridge that was never built. / A map which shows the bridge that collapsed; the streets that never existed'. The general rule is given in the fourth line, where the cartographic freezing of time constantly becomes inaccurate in a temporal continuum: 'Today's plan is already yesterday's – the streets that were there are gone' (*Belfast Confetti* 11). This theme is repeated throughout *Belfast Confetti*, perhaps most explicitly in 'Question Time' where it is stated that 'Belfast is changing daily', that '[m]aps and street directories are suspect' because 'they avoid the moment' (57–8). The representational relationship is complicated rather than transparent in the paradoxical statement concluding 'Revised Version' that '[t]he city is a map of the city' (69). A further inaccuracy, especially relevant in Troubles-era Belfast, is felt in the map's failure to record the shape of prisons 'for security reasons' (11). Carson's poetic project is, in part, to draw attention to these lacunae in the map and to fill in the gaps that they leave. 'Turn Again', the poem's title, emphasizing the interaction between the spatial ('turn') and the temporal ('again'), suggests repetition *not* as an affirmative practice corroborating the already seen, but as the opening up of a space for reassessment. Following a characteristic swerve towards personal experience in the second stanza – 'When someone asks me where I live, I remember where I used to live' – the final line states that 'history has changed' (11). The poem, as an epigraph, suggests that *Belfast Confetti* is just such a textual space for reassessment in which to 'think again'.

The inability of cartography to represent change seen in this poem is given two different weightings throughout Carson's work. Firstly, the map's authority is deconstructed: indeterminacy is offered as a mode of resisting the binary logics of the Troubles which define allegiances along an impermeable Nationalist/Loyalist sectarian fault line. Indeterminacy also undermines the military's strategic employment of the map as a fixing of the present in its attempt to contain the situation. Secondly, the limitations of maps and their reference to a single moment in time offer the map-reader access to a past experience of place now otherwise lost due to change in the material fabric of the city. These two registers, negative and positive respectively, coexist in much of Carson's poetry.

The second, positive register of the map is elaborated in Carson's introductory note to a reissue of the 1931 Ordnance Survey map of West Belfast: '[m]aps are another way of looking at the world we inhabit; and everyone, looking at a map, will have their own story about that particular corner of the universe. Maps can sometimes tell us where we are; they can help us to remember' (*West Belfast* n.p.). Furthermore, in *Last Night's Fun* (1997), Carson notes that the Belfast of his dreams is one in which 'the vestiges of antique maps become the map' (33). 'Vestiges', traces of the past in outmoded maps, are in tension with the gaps (such as 'the shape of the jails') to be found in their current equivalents. In both instances, the obsolete map enters into a relationship with its contemporary counterpart, disrupting the latter's authority and initiating personal narratives of memory. In 'Turn Again', the simple listing of long-disappeared or never-realized street names in line three contributes a similar undermining of current order. However, the effect is not entirely positive: old maps in the poem record the loss of an idealized, unscarred Belfast. The Falls Road area of the city referred to in the poem is the Catholic neighbourhood in which Carson grew up and that was subsequently subjected to extensive alteration, its character redrawn at the hands of both military and paramilitary forces, as well as by redevelopers. The deterioration of the 'linen backing' of the map (the means whereby posters or prints are preserved or restored) in the second stanza of the poem, provides a metaphor for the transformation of this area and its 'authentic' character in redevelopment. In short, the map registers in Carson's poetry as a tool for the imposition of authority and is challenged as such, but also constitutes an *aide-memoire*, a spur to memory as an alternate rendering of place. Indeterminacy and recuperation are thus established in 'Turn Again' as the two poles between which Carson's poetics oscillates. The map embodies a tension between critique and consolation in Carson's work.

Cartography is only one format for representing the city amongst many. Maps constitute the dominant form of representation that the authorities attempt to enforce and maintain, whilst various other groups postulate other representations that challenge this hegemony – the poems record and intervene in this competition. 'Carson's aesthetic', Wheatley finds, 'forgoes any illusions of the lyric poem as a sphere of privileged transcendence or exemption from the signifying chaos of the contemporary world. On the contrary, he exults in it' ('Pushed next to nothing' 52). Belfast, as described in *The Star Factory*, is marked by a sectarian geography of peace walls, estates, and enclaves, that is most clearly indicated in the numerous churches' 'neo-Gothic campaniles ... ubiquitous as factories or barracks', as well as by the industrial and military presence with which the churches and cathedrals are interspersed (6). In another passage from Carson's *The Star Factory*, '[h]igh brick walls of overlooking factories advertise their wares of vitriol and linen in fading white-painted letters, and the gable end of a gospel hall commands us to meet our God' (85). Alongside these sanctioned signs in the environment, the poetry intersperses samples of the sectarian graffiti covering Belfast – '*The Provos Are Fighting For You. Remember It. Brits Out*' (*Belfast Confetti* 33), '*Ulster Says No* scrawled on the side of the power-block' (*Irish for No* 49) – a more ephemeral marking of the city. Consequently, for Carson, 'it seems that every inch of Belfast has been written-on, erased, and written-on again' (*Belfast Confetti* 52). In a further development of writing on the city, in at least one instance Carson considers the buildings constituting Belfast, the infrastructure itself as opposed to the writing imposed upon it, as a 'ubiquitous dense graffiti' on the landscape (*Belfast Confetti* 81). The city, then, is a palimpsest, an overwritten space in which different ideologically opposed groups compete to mark both the fabric of the city itself, and to represent it in text and image.

In this contestation, the map is most consistently associated with military control and coercion, elevated perspective, and the panoptic knowledge that this point of view purports to realize and exploit: 'if there is an ideal map, which shows the city as it is, it may exist in the eye of that helicopter ratcheting overhead, its searchlight fingering and scanning the micro-chip deviations' (*Belfast Confetti* 58). This perspective, monitoring 'deviations', signifies an apparently omniscient knowledge that figures in Carson's rendering (in the aptly titled 'Intelligence') as an attempt to 'write' the city: 'this helicopter chainsaws overhead, ... its pencil-beam of light flick[ing] through the card index' (*Belfast Confetti* 78). The act of cataloguing is different in tone and scope to the Sebaldian curation discussed in Chapter 1. The searchlight-as-pencil metaphor suggests the military aim of forcing the city to conform to legibility. In these instances, Carson critiques the military map, a particular sub-genre of cartography, that is not redeemed by the recuperative potential possessed by the other (civilian) maps that appear elsewhere in his texts.

Paramilitary policing of space is the subject of 'Question Time'. Carson's speaker relates a bicycle ride in which he retraces a childhood journey from his home in the Catholic Falls Road area to the library in the adjacent Protestant district of Shankill. The military imposition of a Peace Line – 'a thirty-foot-high wall scrawled with graffiti, mounted with drab corrugated iron' – leads to the narrator's disorientation as the thoroughfares familiar from his childhood are now dead ends (*Belfast Confetti* 60). He eventually finds his way by a circuitous route back to the Falls Road but is then accosted, detained and questioned at length by militant Republicans who regard his movement across the sectarian divide with suspicion. The narrator is interrogated as to the locations of demolished streets and buildings as a means of ascertaining his credentials as someone who 'belongs' there. This enforcement of denominational and spatial identity is specifically associated with the map: 'The map is pieced together bit by bit. I am this map which they examine, checking it for error, hesitation, accuracy; a map which no longer refers to the present world, but to history, these vanished streets; a map which is a moment, this interrogation, my replies' (63). Here, the coalescence of representational forms and practised experience is marked. Like his movement between separated zones, the speaker's knowledge of the area's former (pre-Troubles) layout and character – his mind map of the neighbourhood – as opposed to its current manifestation offers an implicit challenge to cartographic knowledge and the spatialized identity that it enforces. The speaker's identity, incorporating the memory that prompts his journey, exceeds the subject-position in which authority seeks to situate it. Temporality exceeds cartographic attempts to write the sectarian geography of Belfast and impose stasis on the temporal and spatial flux of the city.

Resistance to military legibility is also present in Carson's poems in the form of erasure, and of alternative rewritings undertaken by other (usually paramilitary) groups, whose exact identity often remains unclear. Metaphors associated with writing and script proliferate in Carson's descriptions here as well. In the opening lines of 'Punctuation', cartographic knowledge is inscribed and then undermined: 'This frosty night is jittering with lines and angles, invisible trajectories: / Crackly, chalky diagrams in geometry, rubbed out the instant they're sketched, / But lingering in the head' (*Belfast Confetti* 64). The complexity of that which is mapped – frost patterns, but also the movements of terrorists or suspects, and of the poem's narrator – exceeds diagrammatic and geometric rendering of space. These forms of representation cannot account for trajectories that elude detection and remain 'invisible', and, as the poem goes on to note, '[f]or the moment, everything is *X*, a blank not yet filled in'. Shots are heard but cannot be located or accounted for, and the poem concludes when 'another shadow steps out from behind the hedge, going, dot, dot, dot, dot, dot …' (64). These dots refer back to the poem's title, they indicate the grammatical marker for elision and therefore more gaps or lacunae, but most importantly, they also signify gunfire – a very different form of punctuation. The same metaphorical coalescence of grammatical markers and terrorist violence is found in 'Barfly' where a shooting in a pub is described: 'two punters walk in, / Produce these rods, and punctuate the lunchtime menu'.

The narrator notes that this act 'was a message. Or an audio-visual aid' (*Belfast Confetti* 55). The actions of paramilitary groups are a language with signifying capacities. Carson does not simply endorse these alternate inscriptions; rather, the combined presence in his poetry of multiple representative strategies for marking space employed by different parties undermines the claims to omniscience of any single perspective and representation (cartographic or otherwise) in the depiction of Belfast. By these means, the poems record the production of the city's spaces by a process of overdetermination: the markings of different groupings are co-present upon the fabric of the city, and in Carson's own composite representations thereof. The space of the text mobilizes multi-perspectivism against military and paramilitary unilateralism.

The absolute authority of any single written version of Belfast is further undermined by the subjective perception of the city provided by Carson's narrators. That is, the uncertainties of reading processes are added to those of writing. Mappings are thus doubly indeterminate. In 'Gate', the signs seen in the landscape fail to communicate and doubt becomes the central register of the poem: 'Passing *Terminus* boutique the other day, I see it's got a bit of flak: / The *T* and the *r* are missing, leaving *e minus*, and a sign saying, MONSTER / CLOSING DOWN SALE' (*Belfast Confetti* 45). The poem's first stanza establishes the Belfast landscape as one marked, or written upon with the traces of the Troubles: the flak from an explosion literally disrupts the legibility of the city as a signifying space. That this space is already written upon by heterogeneous authors is indicated in the proliferation of shop names and advertising. It is also significant, in this context, that the building housing a local newspaper (literally a written manifestation of the city) is damaged in the explosion. The same or similar bomb damage features again in 'News', where '*The Belfast Telegraph* / sign reads / *fast rap*' (*Breaking News* 17). Different forces compete, within 'Gate', for dominance in processes of representation, but the reconstruction of meaning is shown here to be dependent on the narrator's personal knowledge of the area. This effect is emphasized in both the recourse to popular idiom and in the foregrounding of oral narratives as the source for much of this information.

In the second stanza, the explosion, the incident at the heart of the poem, emerges as a central focus, but then immediately loses its distinct identity with the introduction of the narrator's uncertainty regarding the correct sequence of events.

> The stopped clock of *The Belfast Telegraph* seems to indicate the time
> Of the explosion – or was that last week's? Difficult to keep track:
> Everything's a bit askew, like the twisted pickets of the security gate, the
>     wreaths
> That approximate the spot where I'm told the night patrol went through.
> (*Belfast Confetti* 45)

In other poems, the self-perpetuating logic of the Troubles is established – 'the present is a tit-for-tat campaign, exchanging *now* for *then*' (*Belfast Confetti* 30) – but in 'Gate', the narrator's provisional perspective undermines this justification for the cycle of violence. The 'skewing' of chronology finds its register in the environment both in the stopped clock – a clear indication of temporal disruption – and the 'twisted pickets of the security gate'. The proviso that it is 'difficult to keep track' and the fact that the wreaths only 'approximate' the site where events took place are further spatialized marks of indeterminacy. As Alex Houen observes, Carson's writing is 'saturated with the conditional and the subjunctive' (261).

The poem's first word – 'passing' – signifies both the centrality of the narrator's movement and perspective in the generation of the poem, and the temporal theme itself. Memory and perspective are seen to intervene in the recollection of events reported *in medias res* (emphasized in a trochaic opening rhythm) and therefore in the formation of the city's representation. The writing of Belfast's history and the production of its spaces are open to ongoing processes of remaking and reinterpretation: the model of representation invoked is one that, as the poem 'Hairline Crack' describes it, holds 'Everything Provisional' (*Belfast Confetti* 50), or later again, in 'Peace', where 'Everything was, as it were, provisional' (*For All We Know* 55). Carson is also undoubtedly aware of the other more established associations of the word 'provisional' in the context of the Northern Irish dispute (the Provisional IRA, etc.). His point is, in part, to (re)introduce polyvalence in the term. The poetry grapples with the problem of writing indeterminacy that is faced by environmental or spatial texts. 'Gate' is paradigmatic of the strategies Carson deploys for engagement with the cartographic representations of Belfast throughout his poetry. It undertakes a dual project, aiming to interrogate and undermine the existing signifying spaces written during the Troubles and since; and, perhaps more importantly, reconceiving the production of spaces and histories not as already-produced but as in a state of flux and constantly subject to the process of remaking. The discipline of cartography is undermined by the constantly changing environment that it aims to consolidate.

Carson's poetry employs a range of narrative techniques in the undermining of the claims of singularity and universality made for particular representations of the city. Numerous shorter poems – 'Barfly', for example – are marked by distinguishable and colloquial registers that serve to differentiate their narrator from the authorial persona uttering other poems. Additionally, several longer poems shift between registers to present a composite multi-perspective amalgamation of viewpoints and the versions of the city that they produce. The long-line meter that Carson employs throughout both *The Irish for No* and *Belfast Confetti* emphasizes a conversational tone and contributes to the characterization of narrators. Poems such as 'Queen's Gambit' establish various situated perspectives on a series of events, dispersing focus in the space of the poem both *topographically* in the landscape that it describes and *typographically* upon the page. Emphasis is laid on perspectival shifts and on the mediation of each attempt to represent or mark space by another. As Alan Gillis describes it, 'the style of the telling hugely complicate[s] the tale, insistently unveiling the layers of formal mediation that always come between ourselves and history' (193). The short lines of *Breaking News* are formally distinct from these earlier collections, but Wheatley notes that they too 'embrace the provisional, the gapped and the fragmentary over a measured and leisurely utterance' ('Pushed next to nothing' 51).

In Carson's poetry, alternative apprehensions of place are often predicated upon identities exceeding simple sectarian definition. Again, memory and personal history are crucial in the establishment of these narratives running alongside and counter to cartographic modes of understanding. 'Ambition', a poem in which memories of his father constantly intervene in the narrator's perception of current events, articulates and simultaneously subverts a 'time as a road' metaphor:

> Though often you take one step forward, two steps back. For if time is a road,
> It's fraught with ramps and dog-legs, switchbacks and spaghetti; here and there,
> The dual carriageway becomes a one-track, backward mind. And bits of the landscape
> Keep recurring.
> (*Belfast Confetti* 27–8)

Once again, the stasis of the map is undercut by the invocation of the temporal (here as memory); but in this passage an alternate conception of space is offered alongside this deconstructive activity. The landscape is apprehended through recurrence and resonance in a non-linear temporality. The oppositional relationship between cartographic disciplines and this practical 'knowledge' of place is made explicit in a characteristically hesitant statement made later in the same poem: 'I think I'm starting, now, / To know the street map with my feet, just like my father' (30). This more personal register of place relies upon the mobilization of experiential engagement ('with my feet') *alongside* memory and personal history ('just like my father'), in contradistinction to the distanced, panoptic and supposedly objective perspective to which the cartographer's printed map aspires. Knowing the street map by foot also recalls much earlier surveying practices – pacing out – that the printed map is indebted to but also elides.

Furthermore, the establishment of particularized positions of narration is the basis for the inclusion of perspectives occluded from the map. Carson's recollection of his childhood experience of the city returns numerous times to the discovery of alternate perspectives: 'For, although much of the world was withheld from us, we had premonitions of its depth and breadth from these occasional insights: squinting through the eye-knot of a creosoted pine fence, for instance'; and '[s]ometimes the gates in factory porticos were opened to admit a Clydesdale-drawn cart, and the inner courtyard was revealed' (*Star Factory* 2–3). In 'Clearance', the discovery of unknown perspectives combines with the alteration of the city's fabric. The poem reports on a new line of vision opened up as '[t]he Royal Avenue Hotel collapses under the breaker's pendulum': 'There, through a gap in the rubble, a greengrocer's shop / I'd never noticed until now. Or had I passed it yesterday?' (*Irish for No* 32). Again, the narrator's 'passing', on foot, is associated with testifying to the environment passing away. The breaker's 'pendulum' suggests that of a clock and thus also evokes the passage of time. A fresh perspective, opened up by the changing city, produces the same tone of uncertainty and doubt as in 'Gate'. The same double register attaching both positive and negative connotations to the changing city is present in this poem as is discussed above in relation to 'Turn Again'. The 1930s interior of the partially demolished hotel produces the exclamation 'happy days!' in relation to the pre-Troubles era thus recalled (though perhaps ironically as the time recalled is one of economic depression and mass unemployment in Belfast, as in other industrial cities). The poem simultaneously revels in the city's indeterminacy and mourns the loss of its familiar landmarks.

Whilst this nostalgic tone is certainly a strong presence in Carson's response to the changing city, the register that celebrates the proliferation of possible meanings and representations of space in defiance of the map's proposition of a singular authoritative view remains dominant. Several prose pieces dwell upon this potential. 'Brick' describes the city's building and growth: 'Belfast is built on *sleech* – alluvial or tidal muck – and is built *of* sleech, metamorphosed into brick, the city consuming its source as the brickfields themselves were built upon' (*Belfast Confetti* 72). Even in detailing these processes, Carson simultaneously works to undermine the 'proverbial solidity' of the brick in noting its origin in sleech – an 'indeterminate slobbery semi-fluid' (72). The process of expansion is found to harbour within itself the necessity of change, including demolition: 'the very city recycled itself and disassembled buildings – churches, air-raid shelters, haberdashers, pawnshops – were poured into the sleech of the lough shore to make new land'. In addition, the solidity of the brick also suits it to destructive purposes: '[t]he subversive half-brick, conveniently hand-sized, is an essential ingredient of the ammunition known as "Belfast Confetti", and has been tried and tested by generations of rioters' (72–3). Instability inheres not only in perspective but in the city's fabric too.

## The Labyrinthine City

If the panoptic perspective of the printed map is discredited in Carson's texts, the labyrinth is often postulated as an alternative model for a dynamic, open rendering of space. The labyrinth is an ambivalent figure, signalling disorientation and incomprehensibility, but also intricacy and artistry. The negative associations of the spatial form are not absent from Carson's usage: the anxieties commonly attached to being 'lost' are present in his writing. In 'Question Time', the limits to knowledge of place are recognized: '*I know this place like the back of my hand* – except who really knows how many hairs there are, how many freckles?' (*Belfast Confetti* 57). The same root simile is present in '33333': 'I know this place like the back of my hand, except / My hand is cut off at the wrist' (*Irish for No* 39). However, these common connotations of the labyrinth are not dominant. Rather the labyrinth is configured in a complex interaction with the map and often presented as a more viable model. Both of the instances quoted here explore modes of knowledge and undercut certainty. In the context of the Troubles, a version of mapping that exceeds explanation and resolution within the confines of existing modes of knowledge signifies a potential to which Carson is drawn. In short, the labyrinth – incorporating the view from above and the view from within – makes for an indeterminate mapping of space in contradistinction to the fixed, panoptic perspective inscribed by the map.

However, in Carson's poetry, the relationship between the two is not one of simple opposition. A number of trajectories are established. In the unpublished essay 'Belfast', Carson describes the 'dense, herringboned street[s]' of the Lower Falls area of his childhood as a 'labyrinth of spinning mills and foundries'. This 'townscape', imagined as a woven fabric ('herringboned'), 'remained unchanged in its essentials until the early Seventies, until all was swept away in a rash of redevelopment' ('Belfast' n.p.). This redevelopment is represented as a loss of the authentic: 'All this has gone, of course, and I hardly recognize the Falls Road now: the labyrinth has gone, with all its nexuses of eccentricity; whole important urban nooks and crannies, venues of social interface, have been wiped off the map' (n.p.). The trajectory of official redevelopment is one moving from 'labyrinth' towards 'map' (and erasing the former). Carson's poetry of place constantly seeks to disrupt and reverse this process – it is the labyrinthine character of place that is upheld in his writing. In this passage, the description of redevelopment lays emphasis on its rationalizing tendency in the erasure of 'eccentricities' of environment, and of 'nooks and crannies'. The process described is one of streamlining. By contrast, Carson's terms of description explicitly load those characteristics that are removed with value and importance: they constitute the 'nexuses' of community, the 'venue of social interface'. The same trajectories are present in 'Intelligence'. In opposition to the tendency for removal, simplification, knowability, uniformity, Carson's texts re-inscribe that which is removed. In *The Star Factory*, the flipside to the elevated perspective of the map is drawn out in a rare description of subterranean Belfast: 'one ponders the alternative hologram of the city described by its ubiquity of plumbing and its labyrinthine sewers, the underworld of culverts plunged in stygian gloom' (2). The labyrinth offers the alterity that both the sectarian divides in Belfast and the military management of the situation attempt to elide. In this instance, the truly labyrinthine character of the city is seen to underlie its superficially 'mapped' and charted surface. Entrenched positions are confronted in recourse to the labyrinth.

Both the map and the labyrinth are present in 'Belfast Confetti'. The first of the poem's two stanzas plots the riot and explosion constituting its subject matter on 'the map': 'Suddenly as the riot squad moved in, it was raining exclamation marks, / Nuts, bolts, nails, car-keys. A fount of broken type. And the explosion / Itself – an asterisk on the map. This hyphenated line, a burst of rapid fire …' (*Irish for No* 31). The map, the dominant apprehension of space in this stanza, is connected to the attempts to write or mark the environment on the part of different groups in the same manner as described in other poems discussed above. Linguistic and grammatical markers provide an extended metaphor for the description of the scenario: missiles thrown by rioters (the 'Belfast confetti' of the title) are likened to exclamation marks, military attempts to control the situation (and the space) mark the explosion as 'an asterisk on the map' and block up exit routes with 'stops and colons', the 'rapid gunfire' is immediately followed at the end of line three by an ellipsis signifying the sound of the repeated discharge. Marking the environment itself and representing it in text and image are conflated, but any semblance of verisimilitude is problematic in the employment of 'broken type'. A constricted geographical environment is translated into representational difficulty, which is then replicated in the poem's irregular occupation of the space on the page with a shifting metre, long line length, and sustained enjambment. In short, the first stanza works, like numerous other poems, to undermine the signifying capabilities of cartography.

However, in the second stanza the labyrinth is substituted for the map. The transition from one to the other in this second phase of the poem is emphasized by the stanza break itself and by the shift from past- to present-tense narration:

> I know this labyrinth so well – Balaclava, Raglan, Inkerman, Odessa Street –
> Why can't I escape? Every move is punctuated. Crimea Street. Dead end
> again.
> A Saracen. Kremlin-2 mesh. Makralon face-shields. Walkie-talkies. What is
> My name? Where am I coming from? Where am I going? A fusillade of
> question-marks.
> (*Irish for No* 31)

Two variant geographical perspectives compete with one another in this single poem. The type of knowledge that the re-orientation towards the figure of the labyrinth inaugurates ('I know this labyrinth so well') is distinct from that of the poem's opening and from the military enforcement with which it closes. The seemingly oxymoronic claim – 'knowing' the labyrinth (the spatial figure of confusion) – requires a reconfiguration of knowledge itself away from panopticism, and towards knowing 'by foot' from the street. Here, Carson is drawing on what Peter Smith has described as the 'maze factor' where the labyrinth is 'a sign of citizenship in cities; to know its secret confer[s] both status and the freedom enjoyed by burghers' (171). This explicitly civilian form of 'knowing' contrasts sharply with that of military intelligence. Indeed, Tim Ingold's distinction between 'inhabitant knowledge' and 'occupant knowledge', in which walking is associated with the former, translates onto this civilian/military dichotomy (*Lines* 88–9).

In this, Carson's texts also clearly engage with ideas that have been formulated by Michel de Certeau. The antagonism that de Certeau identifies between disciplinary procedures governing society – 'strategies' – and individuals' ways of using imposed systems – 'tactics' – has been influential in research pertaining to landscape, space, and place, and especially in literary geography (34–9). Indeed, de Certeau's own study plays out the spatial ramifications of this tension in some detail. Of particular relevance here is the distinction he draws between the tendency for elevated perspectives to 'immobilize' the city's 'opaque mobility' whilst 'the ordinary practitioners of the city live "down below", below the thresholds at which visibility begins' (92–3); or the observation that whilst maps 'form tables of legible results', 'stories about space exhibit on the contrary the operations that allow it' (121). In these terms, Carson's use of the labyrinth approximates the latter over and above the former in each of these pairs. However, it is not my intention here to trace the resonances that reverberate between Carson's and de Certeau's ideas (several critics have already undertaken this kind of reading), but to elaborate some of the instructive ways in which the two differ (Alexander, *Ciaran Carson* 95–8; Goodby).

Critiques of de Certeau's work have focused on the strong distinction drawn between strategies and tactics: panoptic vistas are associated only with official viewpoints and systems of control, and are condemned as such, whereas street-level, individuated, tactical perspectives are unquestioningly celebrated. David Pinder critiques the 'romanticized version of "resistance"' in de Certeau's work, and observes that his 'dichotomous account, with its advocacy of the "view from below" in opposition to the "high view", can settle into an unhelpful fixed antagonism with its own simplifications and distortions' (401, 403). For Massey, de Certeau's dichotomy 'involves a conception of power in society as a monolithic order on the one hand and the tactics of the weak on the other'. In the process it overestimates the coherence of the powerful and overlooks the implication of the weak in power (*For Space* 45). For Alexander, 'Carson's writing of Belfast notably complicates de Certeau's account of walking in the city' (*Ciaran Carson* 107). In a similar vein, Goodby notes that in Carson's work, 'the oppositions are not simple ones' (77). Personal labyrinthine knowledge is not simply endorsed in place of disciplinary cartographic knowledge in 'Belfast Confetti'. Rather, the two are shown to be mutually implicated. The narrator's knowledge does not overcome the military 'punctuation' of the environment – his 'path-finding' fails, he cannot escape the situation and is subjected to the military questioning at the poem's conclusion.[2] Furthermore, the maze-like quality of the environment is in this instance at least partly the result of the military's erection of blockades. The space of the poem itself echoes the environment described in its heavily punctuated state. This is felt particularly in the second stanza where caesurae proliferate and no clause exceeds six words without the incursion of a grammatical marker of some kind (commas, full stops, dashes, question marks). Brosseau has suggested that whilst de Certeau and others are 'very sensitive to the practices of everyday space, their reading remains largely sociological and their writing seeks to communicate this experience transitively', whilst literary texts 'enact it through rhetorical figures and an active work on syntax' ('City in textual form' 101). Carson's syntactically circumscribed poem certainly functions in this way. If, as Gilbert suggests, '[w]alk poets are … able to evoke in the very spacing of the verse the rhythm of a trajectory punctuated by choice, pausing in one place or leaping ahead as the moment dictates', then Carson's constrained verse echoes the restricted walking that it describes (27).

---

2 If 'Belfast Confetti' describes civilian entrapment, then 'Trap' provides an intertextual counterpoint in which a soldier is ambushed (*Breaking News* 15).

In 'Belfast Confetti', the labyrinth appears more apposite than the map in describing an event in place. If the poem does not provide a clear way out, it also refuses closure and inaugurates a new mode of resistance therein. The military's 'fusillade of question-marks', with which it concludes, open out rather than secure positions and identities. The role played by geography in identity formation is certainly at issue in this poem, and the function of modes of representation (including mapping) in these processes is interrogated by the intervening set of concerns that inhere in labyrinthine descriptions of space. Meaning remains indeterminate.

The labyrinth, as a literary symbol, is underwritten by indeterminacy. Penelope Reed Doob's study of the origins and later uses of the labyrinth evaluates the two perspectives that it inaugurates and combines:

> [Labyrinths] presume a double perspective: maze-treaders, whose vision ahead and behind is severely constricted and fragmented, suffer confusion, whereas maze-viewers who see the pattern whole, from above or in a diagram, are dazzled by its complex artistry. What you see depends on where you stand, and thus, at one and the same time, labyrinths are single (there is one physical structure) and double: they simultaneously incorporate order and disorder, clarity and confusion, unity and multiplicity, artistry and chaos. (1)

As a result, '[o]ur perception of labyrinths is … intrinsically unstable', and, as images, they are 'convertible and relative' (1). In Carson's hands, the labyrinth becomes the figure whereby the importance of positioned, situated perspective (both spatially and temporally) is emphasized. Furthermore, its use allows for interplay between elevated ('maze-viewing') and street-level or path-finding ('maze-treading') perspectives rather than enforcing a binary separation of the two. In contrast, the map cannot encompass the latter of these two perspectives. In these observations, my argument is differentiated from those put forward by others who have offered extended treatments of the labyrinth motif in Carson's texts. Kathleen McCracken has written that 'the labyrinth, like most of his other dominant images, is flexible and may carry slightly different connotations in each context' (357). Conor McCarthy, developing McCracken's work, details 'a number of ways of representing Belfast' realized in the figure of the labyrinth (115). My own argument suggests that rather than introducing 'slightly different' but essentially similar connotations, the figure of the labyrinth incorporates two antithetical perspectives.

Carson's writing constantly emphasizes the intricate and extensive web of connections between elevated and street-level perspectives. The former are often predicated upon the panoramic views of Belfast made available by climbing Black Mountain overlooking the city. Numerous poems and passages of prose recall Carson's climbing this mountain with his father both as a child and since. Whilst this elevated viewpoint often inaugurates poems, it hardly ever goes unmediated: variation of situation is commonly enacted by the poem's close. 'Patchwork' provides an example of how this intersection functions, opening with an elevated prospect – a way of seeing landscape freighted with a long history of its own (see Cosgrove). It is described, tellingly, as map-like:

> Nearly at the summit, we could see the map of Belfast. My father stopped
> For a cigarette, and pointed out the landmarks: Gallaher's tobacco factory,
> Clonard Monastery, the invisible speck of our house, lost in all the rows
> And terraces and furrows, like this one sheep that's strayed into the rags
> And bandages that flock the holy well. (*Irish for No* 59)

The poem's conclusion, however, gravitates towards a reversal of its opening panorama. The narrator looks up to the mountain from the city: 'far away, up on the mountain, a cloud of sheep had scattered / In the Hatchet Field' (63). The two perspectives interpenetrate each other within a single poem. As Stainer notes, 'the mountain-top observer of the city is simultaneously observed. This form of doubled geographical awareness allows the individual to step outside of the normal, bounded field of understanding' (391). The spare poetics of *Breaking News* compress the same telescoping process, as in 'Home' where 'from the edge / of the plateau / my eye zooms // into the clarity / of Belfast / streets' (12). In *The Star Factory*, similar outings to Black Mountain are described several times. Looking down on 'the maze of factories, mills, barracks, schools, the filing-systems of terraced houses', one can see the many bodies of water throughout the city. The elevated perspective is, however, then modulated when the memory of other street-level perspectives intervenes:

> Wandering at ground-level within the dense urban fabric of brick walls, in the valleys of shadow cast by the tall factories with their blanked-out windows, it was beautiful to get, through the iron rails of a locked factory gate, a glimpse of a wind-rippled mill-dam on which drifted a flotilla of swans. (45)

The same factories perceived from the elevated position are now re-envisioned as the limitations on lines of vision from '*within* the dense urban fabric'. Repeatedly, distanced and inhabiting perspectives are alternately the focus of Carson's texts and, as here, are placed in close proximity and juxtapositional relationship to one another. The potential for one to interrogate the other is realized recurrently. Again, the result is the exposure of indeterminacy inherent in single perspectives. Ultimately, in structural terms, Carson asserts the necessity of multiple and mutually dependent positions of narration.

A further passage from *The Star Factory* describes another, or perhaps the same, ascent of Black Mountain again. From the 'celestial summit', the narrator and his father could look down:

> towards the various details of the urban map spread out below: the biggest shipyard in the world; the biggest ropeworks; the green cupola of the City Hall; the biggest linen mill; Clonard Monastery; Gallaher's tobacco factory; my school; the GPO; and in between, the internecine, regimental terraces of houses and the sprawled, city-wide Armada of tall mill funnels writing diagonals of smoke across the telescopic clarity of our vision. I could see the colour-coded kerbs of Union Jack and Tricolour. I saw my tiny self appearing at the front door of 100 Raglan Street … (166–7)

As in the examples discussed above, the same modulation of perspective occurs here, with the concomitant shift from perception to imagination in the vision of 'my tiny self' outside the family home. Imagination is also present in the seeking out of 'our house' in the passage from 'Patchwork', even though in reality it is an 'invisible speck'. In two of the three descriptions, the elevated perspective is explicitly likened to that of a map and, initially, dwells on public buildings and the social concerns that they represent, before narrowing down to personal narratives in the imagined reproduction of street-level perspective. What is crucial to all these descriptions, however, is that macroscopic and microscopic perspectives thoroughly intersect one another, often actually producing imagined recuperations of their opposite. If elsewhere elevated perspective is associated only with the authoritarian strategies of disciplinary control, the sense here is more democratic. In these passages, tactical uses are insinuated in the prospect, shifting the terms of ownership and power with which it is associated.

In some instances, the same process of subjection to alterity is practised on maps themselves. In an essay accompanying the reissue of the 1931 West Belfast Ordnance Survey map, Carson notes:

> Much of the eastern portion of this map … remained unchanged until the early Seventies; and looking at it, I find myself about the age of five, walking from my house in Raglan Street, along the Falls, turning at the Library (I remember its tang of rexine bindings and floor-polish), up Sevastapol Street, Odessa Street, Clonard Street, then down Waterville Street to St Gall's School. (*West Belfast* n.p.)

A connected set of shifts occurs here: Carson's description modulates, within a single sentence, from looking to walking, from the visual to the olfactory, and – most importantly – from detailing the map's dimensions towards a restoration of the personal narratives that it excludes (but also provokes). This dual perspectival register is labyrinthine.

Whilst the recuperative aspect of elevated perspectives is undoubtedly present in Carson's texts (for example, in the production of narratives of memory from old maps), this is balanced against the potentially destructive charge that those perspectives carry. At the conclusion of *The Star Factory*, the narrator remembers a daydream prompted by the depiction of the Angel Victor flying over various locations on old Northern Irish stamps:

> I indulged a momentary fantasy of Victor over Clonard Monastery, encompassing the whole of Belfast with his carrier-pigeon's aquiline angelic wings, recording all its architecture in the pupil-black orbs of his aerial Dresden vision: two-up-one-down terraced houses, shipyards, spinning mills, tobacco manufactories, tram depots, pubs, chapels, churches, ropeworks, barracks, corner shops, arcaded markets, railway stations, graving-docks, cinemas, post offices, and photographic studios, much of it vanished now, into the maw of time, but still remnant in the memory of denizens and pilgrims like myself, and still extant in my father's mind. (290)

This vision is recuperative: it restores (typically for Carson, in list form) that which has fallen 'into the maw of time'. Yet it also suggests violence and destruction in the association of this perspective with that of the bomber over Dresden. This latter referential frame is familiar from 'Dresden', where the potential for overlooking damage on the ground from an elevated bomber's-eye-view is elaborated (*Irish for No* 11–16, esp. 15). The absence of moderating street-level perspective in both these instances perhaps signals the dangers implicit in a movement away from labyrinthine perspective.

Finally, the concept of the labyrinth also receives negative representation itself in some (limited) instances. In one passage, it becomes the figure for the Troubles: 'Catholics have been intimidated from their houses, part of an ongoing head-counting territorial dispute in North Belfast, where many huddles of opposing loyalties rub up against each other in frictions, factions and fractions, a subject too labyrinthine to be entered into here' (*Star Factory* 94). The labyrinth becomes a multivalent figure in Carson's differing uses of it. However, even when negatively charged, this model is still found to represent the overdetermination of Belfast in a more sophisticated way than the map is capable of. The positive or negative register in which the labyrinth appears in Carson's texts is largely dependent on the presence or absence respectively of suitable knowledges for navigation of its intricacies.

## The Labyrinthine Text

According to the Cretan myth, Theseus, son of Aegeus, is amongst a consignment of Athenian youths to be fed to the Minotaur hidden in the labyrinth by Minos. However, Minos's daughter, Ariadne, falls in love with Theseus and provides him with a ball of thread to be used to find his way out of the labyrinth again once he has killed the Minotaur. This ball of woven thread might also provide a way into understanding the ubiquitous presence of weaving in Carson's work. Weaving, alongside a series of similar practices including knitting, threading, and, perhaps most significantly, spinning yarn, is constantly related to narration and story-telling. If space is understood in terms of the labyrinth, its navigation occurs most often by way of the 'woven' text. If one of Carson's practices is the formulation of a narrative of maps, his work is also underpinned by a mapping of narrative. The ideas of the labyrinth and of weaving combine in processes of narrative formation. The activity of writing itself is thought of, in *The Star Factory*, as 'trying to snag a thread which might be spun into a yarn' (with the final word exploiting both the spun thread itself and the rambling story) (76), whilst in the poem 'Second Language', the weaving loom and the typewriter coalesce: 'Bobbins pirn and shuttle in Imperial / Typewriterspeak. I hit the keys. The ribbon-black clunks out the words in serial' (*Ballad of HMS Belfast* 98). The resulting narratives (or 'serials') can be thought of, as in *The Star Factory*, as 'labyrinthine anecdotes' (55).

Doob glosses this historical combination of concepts, noting that '[t]he labyrinth, and the related figure of the interlace, are interwoven, *textus*, like poetry itself.'

Furthermore, '[d]isposition, the ordering of material', is labyrinthine: 'artistic ordering inherently, if not explicitly, resembles the labyrinth, paramount emblem of artificial order' (209, 207). The labyrinthine characteristics that Carson's texts are built around are complexity and polyvalence: narrative is self-perpetuating, forever producing more narrative and generating further associations. In addition, as Carson notes, narration constantly introduces nuance: it is 'always new, because each telling of the story is rehearsal, and gains different subtleties of emphasis each time round' (*Star Factory* 2). These are the qualities of storytelling that make it fertile ground for resistance to the imperatives of the Troubles: in its intricacy and proliferation, narrative necessarily exceeds and therefore contests the entrenched and delimited positions of the sectarian conflict.[3]

It is worth noting here that in using the labyrinth as a figure for both the city and for writing, Carson is drawing on a long tradition. Faris notes that '[t]he labyrinth has persisted from classical through contemporary literature as image and as structural design in urban writing, highlighting the interdependence of those two entities, the city and the text' (33). Furthermore, Steven Winspur suggests that 'city streets function not only as an element of verisimilitude in the overall décor of the modern novel but ... are also a metaphor for narrative itself' (60). Carson alludes clearly to this interpenetration of city and text in his use of a passage from Walter Benjamin as an epigraph to *Belfast Confetti*. The passage suggests that the maze-like city approximates 'the labyrinths on the blotters of my exercise books' (14).

Carson's labyrinthine poetics are comparable to Kristin Ross's description of a poem as constituting 'a movement and not a tableau, a *récit* [story] rather than a map.' Ross goes on to note that '[i]nstead of the abstract visual constructions proper to the stasis of a geographic notion of space, the poem creates a "nonpassive" spatiality – space as a specific form of operations and interactions' (35). In response to a seeming tension between the geographical synchronic and the narrative diachronic, Edward Soja has described the need for a 'viewpoint that pointedly flows from the resonant interplay of temporal succession and spatial simultaneity' (3). Carson's poetics is labyrinthine because it is situated at just such an intersection. As a result, the kind of 'nonpassive spatiality' that Ross describes, allied to the story rather than the map, is produced. John Berger has described this shift as a change in the mode of narration:

> It is scarcely any longer possible to tell a straight story sequentially unfolding in time. And this is because we are too aware of what is continually traversing the story *laterally*. That is to say, instead of being aware of a point as an infinitely small part of a straight line, we are aware of it as an infinitely small part of an infinite number of lines, as the centre of a star of lines. Such awareness is a result of our having to take into account the *simultaneity and extension* of events and possibilities. (40)

---

[3] In suggesting a labyrinthine textual aesthetic in Carson's work, the 'multicursal' labyrinth (where the maze presents different options of direction) is intended rather than the 'unicursal' labyrinth (where one path leads inexorably to and from the centre by a circuitous but single route). Thus, indeterminacy rather than inevitability is suggested.

In Carson's poetry linear progression and lateral intersection occur simultaneously – the former movement is constantly traversed by the latter. Both *The Irish for No* and *Belfast Confetti* progress from one poem to the next sequentially, but both collections are also constructed to emphasize non-linear connectivities. Shifts in levels of narration, metrical complexity, and a series of intertextualities between poems within a collection and across collections, all combine to formulate a truly spatial text. *Breaking News* continues the pattern, most clearly in the dialogues between the separate poems 'Breaking' and 'News' but also by redeployment of techniques from the earlier collections. *For All We Know* takes establishing lateral connections to a new level in its engagement with musical form. The collection emulates the way in which 'the fugue must perform its often stealthy work with shifting // melodic fragments that remain perpetually in / abeyance, or unconsummated' (108). Split into two parts, *For All We Know* repeats a series of identically titled poems, with recurring phrasings throughout. By these means, the text as a whole is a labyrinthine rather than cartographic one.

The labyrinthine character of storytelling in the poems is manifested narratologically in their polyphony. That is, in the marked presence of multiple voices present in variant layers of narration. 'Dresden' is paradigmatic, incorporating a wide variety of the techniques that Carson deploys repeatedly to this end. The poem is replete with markers of verbal relation, and shifts between different levels encompassing its primary narrator, the reportage of Horse's narrative, and the un-attributed presence of numerous minor characters' speech. Inflections and modifications of diction (for example, the inclusion or absence of cliché and popular idiom) mark different narrators. Punctuation – most commonly the mid- or end-line dash – also indicates shifts, whilst emphasis is laid against the replication of the spoken word in the long-line metre and frequent enjambment. These stylistic markers indicate the inter-mixing of narratives that the poem undertakes and the associative structure by which it operates. Association also becomes a thematic focus as the narrator attempts to rein in his tale: 'for this is not the time or the place to go into / Comparisons, and this is really Horse's story, Horse who – now I'm getting / Round to it – flew over Dresden in the war' (*Irish for No* 14). Sub-clauses and dramatic asides constantly punctuate these lines with caesurae: the aerial, cartographic perspective looking down on Dresden (with the same military associations developed at length elsewhere) contrasts with the disassembled space of the text itself. The poem closes with a return to the frame narrative of a visit to Horse's caravan. Here, artificial limits have to be set to the narrative flow of associations in order to reach an end: 'there's no time / To go back to that now; I could hardly, at any rate, pick up the thread. / I wandered out through the steeples of rust, the gate that was a broken bed' (16). The woven, 'threaded' narrative itself 'wanders' through a series of associations that literally exceed this setting of arbitrary confines in their resonance across and beyond the collection in which the poem appears. Significantly, 'Dresden' is the opening poem of *The Irish for No*, and, as such, sets up reading practices for the following poems. The description of piles of rust as 'steeples' perhaps suggests the religious sectarian subject matter through which the poems 'wander' (that is, move without regard for established routes or a particular agenda), whilst transgressing the borders signified by the gate (which is, tellingly, not actually a gate). In short, the poem indicates the capacity for contravention of cartographic rules that narrative of a labyrinthine order creates.

The poetry, then, in its inclusion and negotiation of multiple voices, more regularly conforms to narrative rather than lyric classification, although it seems to work towards a blurring of any distinct boundary between the two.[4] Again, *Breaking News* goes further. Here, the writings of Crimean war correspondent William Howard Russell are appropriated, often silently, with Carson's concluding acknowledgements describing how he has 'taken his words *verbatim*, or have changed them only slightly to accommodate rhyme and rhythm' (74). Borrowing is also especially evident in a run of ekphrastic poems considering paintings by Gericault, Goya, and Hopper, who Carson casts as other species of war correspondent (42, 47, 54). In this company, Carson's own poems are to be thought of as a species of war correspondence for the Troubles. Often, across a number of collections, apparently omniscient positions of narration are revealed, by way of the mid-poem introduction of a frame narrative, to be the reported speech of a situated, unreliable narrator. 'Cocktails' provides a clear example of this repeatedly deployed technique. The first four lines detail a car-chase before, mid-line, the layered narrative is revealed: 'So / The story went: we were in the Whip and Saddle bar of the Europa' (*Irish for No* 41). The narrative structure of the poem exposes determined positions of narration to their own fallibility. The delayed revelation of the positioned narrative voice – that is, positioned both in the cityscape *and* politically – necessitates a revaluation and usually a downgrading of the narrative's authority.[5] In this context, 'Cocktails' addresses the myth-making that contributes to the persistence of the Troubles. The poem's title exploits homonymy to conflate the barroom setting (in which so many of Carson's poems have their genesis) with the submerged presence of the 'tales' told there. This scenario is elaborated in the description of a pub's interior in *The Star Factory*:

> The Crown Bar Box [a variety of snug] is a kind of time-machine, whose interior can accommodate the conversational buzz of about a dozen characters, each with tall anecdotes to tell, whose different times and places interpenetrate the fabric, making a noise like the atmospheric crackle of several short-wave radios yakking simultaneously, hardly interrupted by the giving of a complex order to the bemused waiter. (132)

---

4 Neil Corcoran finds that Carson is 'certain that the lyric tradition needed to be disrupted' and consequently 'constructs a narrative poetry of his own'. The terms in which Corcoran describes Carson's poetics gesture towards the labyrinth without actually using the phrase: 'Carson's is an exfoliating narrative of turnings and returnings, digressions and parentheses, lapses and dissolvings, the haphazard and the circuitous' (214, 216).

5 The same technique is also used on a larger scale in Carson's prose writing, particularly *The Star Factory*. Four pages into a chapter purporting to be omniscient third-person narration, the mask is lifted: 'Such were the stories my father used to tell me' (64).

'Tall anecdotes' intersect one another and produce a cacophony exceeding easy interpretation. This barroom scene, in its repeated deployment, stands in metonymically for Carson's own narratological techniques where 'labyrinthine anecdotes' interpenetrate and interrogate one another. The effect is, again, one of indeterminacy.

If the woven text is valued positively in Carson's writing, the weaving metaphor is also negatively charged in other instances. It is, like the labyrinth, polyvalent in Carson's multiple deployments. 'August 1969', a poem dealing with events regarded, in hindsight, as the opening phase of the Troubles in Belfast, details the burning of Greeves's Mill in some of the earliest riots of the conflict. The significance of focus being devoted to one of Belfast's many *linen* mills – itself a site of weaving – in this representation of historic events is noteworthy. In Carson's description, the significance of these events as the opening of the Troubles-era metaphorically registers as a woven fabric: 'The weave is set: a melt of bobbins, spindles, shuttles' (*Irish for No* 35). The sectarian dispute inaugurated in these events is 'set' and cannot be easily unpicked, but the submerged idea of the urban melting-pot twists the line's meaning in another, more complex direction. 'Smithfield Market' includes another version of the weaving metaphor. The destruction of the covered market and arcades by fire in 1974 is described as 'Everything unstitched, unravelled', leaving only 'mouldy fabric' (*Irish for No* 37). Here, the process of unstitching describes the loss of an environment to alteration comparable to 'Turn Again', where 'the Falls Road hangs by a thread' (*Belfast Confetti* 11). Combining processes of ravelling and unravelling, 'Travellers' describes how 'Belfast / Tore itself apart and patched things up again' (*Irish for No* 42). The repeated but distinct use of these related ideas serves to thread the collection itself together (as well as formulating connections between collections), weaving some narratives and unpicking others. Just as *The Irish for No* opens with the idea of the threaded narrative in 'Dresden' it also loops around to close with it in the final poem, 'Patchwork'. Here, the narrator reports his grandmother's observation on her patchwork quilt that 'The stitches show in everything I've made' (63). The same is true of Carson's poetics of weaving: the stitches by which links are made are emphasized rather than disguised.

The weaving together of thematic and structural links occurs at an inter-collection level as well as within collections at an inter-poem level. A further dimension of the spatial aesthetics of (inter)textuality is thus exploited to extend the potential for resonance beyond the borders of the individual text. Carson's observation, in *The Star Factory*, that a 'hook-and-eye principle is applicable to my own method of writing' (another weaving metaphor), is realized both in *and beyond* each of his texts (226). Repeated use of key phrases in multiple contexts, and hence with multiple meanings, permeate the sequence.

Striking similarities of structure employed in both *The Irish for No* and *Belfast Confetti* also serve to draw thematic threads between the two collections. Both are divided into a tripartite structure where longer, memory-driven associative poems of the first and third sections of each collection are characterized by a conversational, personalized narrator. Part II sees a shift to a more oblique narration (though still largely in first person and frequently encompassing other voices and their storytelling). This textural change is accompanied by the universal employment of a standardized stanzaic poem structure (two stanzas – the first of five lines, the second of four) in Part II of both collections. An alteration in subject matter occurs concomitantly with these divisions and changes: the violence of the sectarian conflict is largely limited to the middle section of both collections. The conflict's presence in part I is oblique, but the first poems of Part II in both collections ('Belfast Confetti' and 'Gate') both open, *in medias res* and initially with a trochaic metre, in the midst of a scenario emphatically indicating the Troubles (a riot and the aftermath of an explosion respectively). Carson thus inscribes but also sets limits on the Troubles in the space of the collections. *Breaking News* frames this struggle in relation to earlier historical precedents. The collection also looks beyond the familiar Falls/Shankill/city centre mise-en-scène, with the opening poem, 'Belfast', drawing a new east/west geography that might serve to disrupt Troubles narratives. The Troubles are undoubtedly present in *For All We Know* – the lovers are brought together by a bomb explosion – but the collection's priorities lie elsewhere. This imposition of borders is in contrast to the constant transgression of limitations that narrative itself enacts.

The thematic oscillation between indeterminacy and recuperation is echoed, in formal terms, in the tension between the fragmentation of narrative voice and the gesture towards an organic form in the shared, woven structure of Carson's collections. Alexander's observation that Carson's aesthetic is 'simultaneously deconstructive and conjunctive' and 'typically dismantles and breaks down its subject matter only to reassemble it in some new and unexpected form' finds poetic expression in the deployment of structural techniques ('Mapping Junkspace' 509). The same tension is played out in the common structure of individual poems in *The Irish for No* and *Belfast Confetti*. Peter Denman notes 'the ghost of the sonnet form' in the nine-line poem uniformly used in Part II of both collections: '[t]he five/four division of these nine-line poems mimics the eight/six division of the sonnet form, and the reduction in the number of lines is compensated for by making each line longer' (32). The poetry metrically navigates between fragmentation and organic form: it is a 'compromise between the free-verse form of the long-line and the formalism of the stanza' (39). If *Breaking News* sees a radical formal departure, then *For All We Know* offers a thoroughgoing resumption of engagement with the sonnet form. In a metaphorical sense, the sequences – especially culturally freighted sonnet sequences – established within and between collections map a narrative. However, the poetic procedures employed also deconstruct that mapping process. Perhaps describing this activity, the concluding lines of 'Hamlet', the final poem of *Belfast Confetti*, find the narrator trying 'to piece together the exploded fragments' (108). That pieces are 'exploded' recalls the bombings of the Troubles that are a thematic refrain throughout the collection, but 'fragments' also indicates a particular relationship to modernist and, in particular, Eliotic poetics. Carson moulds labyrinthine departures from regularly metered verse, but simultaneously recomposes textual space.

## Registering the City

In *The Star Factory*, Carson recalls childhood excursions to the building sites of the new estates being constructed on what was then the periphery of Belfast. At a 'leftover landscape' near Mooreland, Carson would 'watch the bricklayers ply their trade' (122, 126). Their 'trade' is likened to the craft of poetry. Narrative formation is compared to builders' work expanding and redeveloping the city: 'Had their basic modules been alphabet bricks, I could have seen them building lapidary sentences and paragraphs, as the storeyed houses became emboldened by their hyphenated, skyward narrative, and entered the ongoing, fractious epic that is Belfast' (126). In this passage, then, Belfast's expansion and transformation takes place in literary representation as well as materially. The two processes operate in tandem. Textual Belfast is built in an accretive or cumulative manner: contributions (texts) form 'storeys' in houses, and houses parts of the city, texts are 'hyphenated' (linked to one another). A woven narrative of the city emerges in which Carson's own writing can be considered as part of a collaborative project, a contribution to the epic of Belfast. If other groups active in the Troubles compete to write an authoritative, singular version of the city, Carson's texts are to be conceived, antithetically, as a cooperative intervention in an ongoing, open process.

The idea of an individual account embedded within a history of representations appears elsewhere in Carson's writing. In a chapter of *Last Night's Fun* entitled 'The Ravelled Hank of Yarn' – a further deployment of the weaving metaphor suggesting the manufacture of a single thread from multiple fibres – Carson describes the interplay between variation and constancy of 'cheironomy' in musical and choral forms. His comments implicitly elucidate the collaborative role played by his own texts:

> In the cheironomy of the Jewish chant, it should be noted that the single sign hardly ever corresponds to the single note but to an organism of notes; i.e., to an elaborate melodic motif; and that these motifs are never themselves defined sequentially note for note. This means that internally the aggregate of notes remains fluctuating and loose while their character as a group remains constant. … Nearly all ancient oral traditions are surprising for this double feature of endless variety within a fixed framework. (27–8)

As a result, the individual chant 'is connected to a running prose text of greater dimensions' (28). Interplay between variation and fixity (indeterminacy and recuperation, fragmentation and organic form) is a key concept in Carson's depiction of Belfast, as well as in the relation of his writing to a history of representations of that place. In another passage, Carson describes the relationship between musicians (and, by implication, writers as well) to a shared tradition:

> All great musicians recognise their ancestry and pay respect to it, and they know the thing is bigger than the sum of the individuals: it progresses in a multiplicity of exponential steps and fractal variations; and stepping on a butterfly way back there in the past will have an unforeseen chaotic implication for the present or the future. (28)

Art, in the form of music or literature, is conceived as a collaborative and ongoing project in which particular representations contribute to a corpus comprising more than the sum of its parts – a project that elides the notion of a finished or completed version and produces indeterminacy in balance against fixity. The mobilization of indeterminacy that takes place in different forms throughout Carson's texts exposes and questions entrenched positions of allegiance along sectarian divides in Belfast and the definitive maps that they draw.

The means by which Carson's texts realize the fluidity of the city that they represent is, as the foregoing analyses demonstrate, most often by recourse to the subjective viewing process. As Gillis notes, '[t]he view contains the viewer – that there can be no singular, panoptic view of a heterogeneous field is central to Carson's poetics' (186). Emphasizing personal experiences of the city undermines the panoptic aspirations of the map. Carson's landscapes are thus comparable to Sebald's: both writers are at pains to point out the subjectivity, and often the fallibility and limitations, of their recordings. For Carson, this means the constant return to the arena of experiential knowledge: the geography of his texts is the Falls Road area of his childhood. He describes his writing as 'a reverie of West Belfast' whilst other areas of the city 'lay out of sight and out of mind, as [they] did to me for much of my life' (*Star Factory* 189, 240).[6] Even within this specificity, there is often a further narrowing down of focus to the *lower* Falls Road – the network of now-demolished terraced streets surrounding the family home on Raglan Street that Carson knows so well from personal experience. By contrast, the adjacent Shankill Road area features as alien territory in Carson's writing whilst other areas of the city, such as Protestant East Belfast, remain *terra incognita*. In addition to this geographically circumscribed focus, the marking of partiality in Carson's account of the city has implications shared with other case studies making up this monograph.

At stake here, then, is an interrogation of what it means to record the characteristics of place. In an interview with John Brown, Carson speaks of a desire for his writing to simply engage in 'registering what happens', before going on to complicate this seemingly straightforward idea:

> my main aim was … to act as a camera or a tape-recorder, and present things in a kind of edited surreality. … If there's one thing certain about what was or is going on, it's that you don't know the half of it. The official account is only an account, and there are many others. Poetry offers yet another alternative. It asks questions, I think. It asks about the truth which is never black and white. (Qtd. in Brown 148–9)

Objectivity is firmly eschewed. Processes of curation in Sebald's texts resonate here with Carson's concern for recording – for storage of impressions and their subsequent redrafting. The relationship between recording and redrafting in

6 Carson has also said that '[t]he Falls Road and its wee streets were the known world; beyond that lay monsters' (qtd. in Brown 142).

accounts of place, it will be seen, preoccupy all of the writers studied in this monograph. If, in Carson's statement, 'registering' suggests passivity in depicting, then the introduction of 'editing' mediates that immediacy. Even in *Breaking News*, the minimalist poetics and the adoption of a register approximating reportage nonetheless clearly emanate from a subjective narrating persona. This is perhaps the '*art* of the war correspondent' that Carson refers to there (74, my italics). It is out of this commitment that labyrinthine mappings and texts emerge. The same camera metaphor is repeated in another piece – in some ways contrapuntal to this first – where Carson writes that '[t]he eye is not a camera, the eye of memory still less so': 'There is no fixed viewpoint. The eye flits and flickers around all over the place, taking in bits of this and that, weaving in and out, picking, choosing, shuffling, negotiating, building up a picture that is never static, for everything moves through time and space' ('This is what libraries are for' n.p.). Vision, along with other sensory perceptions, weaves together. This visual process is one of mapping, but it does not share with the discipline of cartography a desire to hold in place or to create a kind of 'official account'. Rather, both of these passages emphasize the 'building' of perspectives and the negotiation that this entails. In this, Carson's poetics of place is one aware of the investments inevitably made in any 'registering' of an environment, and one keen to record their presence in the textual fabric of landscape writing. He thus shares with the other writers that I approach a desire to develop a poetics of landscape aware that as it testifies to place it does so from a particular position *and* from within a long tradition.

# Chapter 3
# Walking: Iain Sinclair

If in my study of Ciaran Carson's writing, walking constituted one means amongst many of achieving a mapping of place, here Iain Sinclair's pedestrian poetics rise to the surface. In Carson's poetry, as John Goodby notes, '[t]he persona is almost invariably that of a pedestrian'. Furthermore, Carson's 'means of locomotion' is 'a major factor in shaping the encounter with the world and individual subjectivity, and our readerly encounter with him' (66–7). Indeed, Goodby's findings for Carson offer a concise summation of the chief interests of this chapter: Iain Sinclair's encounter with the world is undoubtedly that of the pedestrian, and the texts that result from that encounter certainly work to articulate the specificity of the pedestrian's perspective in their descriptive aesthetics and narrative style.

'Walking is the best way to explore and exploit the city', Sinclair declares in a passage that critics gravitate towards, as they attempt to define his approach to London (*Lights Out for the Territory* 4). Comparatively few of those critics, however, have noted the apparent tension between ideas of exploration and exploitation in processes of engaging with city spaces that this phrase leaves unresolved. Exploration, alongside cognate concepts of discovery, suggests a process of stepping into the unknown to find what the city offers and, consequently, of adopting a stance open to encounters with the unexpected. Exploitation carries with it the implication of going towards an encounter with a pre-conceived agenda, garnered perhaps from prior research, to be confirmed or (less often) disproved. The written accounts of the city that each of these modes of engagement will produce are different: exploration will generate a record of an experience of place, exploitation will process these details in conjunction with other, prior accounts and representations of place and integrate findings accordingly. In practice, as Sinclair appears to be aware, no description is entirely one or the other but incorporates aspects of both. Yet it remains for the critic to make an assessment of the balance between exploration and exploitation that makes up the writing of place.

The seemingly neat binary between the two ideas is, however, troubled when the etymology of exploration is considered: the term's first English usage is in the sixteenth century, at the dawn of the colonial era, when exploratory journeys were not necessarily devoid of exploitative motivations. As a result, the term's ostensible conferral of agency onto place in the moment of the encounter is thrown into doubt. If Sinclair's texts are often critiqued for their strong reading of place, the embedded methodological statements that permeate them acknowledge this and attempt to guard against it (increasingly in his later work). Thus, the dialogue between exploration and exploitation is later reformulated: 'Watch a street and you become it. You construct, if so inclined, a narrative: but you are also part of the witnessed event. You shape what you see' (*Edge of the Orison* 102). This text casts walking as 'logging evidence', a twofold process made up of 'transcribing' and 'interpreting', wherein recording and describing meet (*Edge of the Orison* 6–7). More recently still, Sinclair has written about stepping back from this very subjective mode of 'witnessing', and instead aiming to 'accept the dictation of place': 'The trick [is] to delete internal projections and fantasies, mental trailers, that act as a nuisance filter between world and self, and to empty that space until landscape flow[s] through, freely and without editorial interference.' Writing of this kind foregoes agency to place, transcribing rather than arranging. Instead of forming 'the armature of a narrative', incidents 'could stand, without embellishment, as a list, a fragment with no beginning and no end' ('Water Walks' 18–20). Here, Sinclair is addressing the same issues with which Carson is engaged when the latter speaks of simply 'registering what happens' (qtd. in Brown 148–9). The issues that closed the previous chapter therefore also open this one. Whilst none of Sinclair's longer texts seriously attempt this kind of account, this chapter argues that his writing loosens narrative bounds to find a representational form that navigates between exploratory and exploitative modes of engagement and is capable of dramatizing the experience of place.

Primary focus is devoted here to *London Orbital* (2002), Sinclair's circumnavigation of London loosely following the M25 motorway, and the text in which the focus on these issues is most sustained and most interestingly inflected. Walking is at the heart of this project. Indeed, as the above quotation indicates, it is the pre-eminent generative process at Sinclair's disposal and thus receives methodological and thematic treatment in many of his other texts, in particular, *Lights Out for the Territory* (1997) and *The Edge of the Orison* (2005), with which *London Orbital* forms a loose trilogy.

## The Syntax of London

In *London Orbital*, as in all his texts, Sinclair elaborates his methodology for representing the city by recalling 'a useful genealogy' of other writers, by which he might 'find some way to unravel the syntax of London' (*London Orbital* 158). The recourse to a literary tradition of representing London is only one part of a larger whole in defining Sinclair's engagement. However, in this instance, when coupled with the conflation of the city's fabric with textual form (as a 'syntax'), the city might seem to be determined by what is written about it rather than how one might experience it directly. If critics have tended to read Sinclair's texts in this way, it is because those texts, in instances such as this, appear to explicitly call for such a response. However, Sinclair's verb is ambiguous – to 'unravel the syntax of London' could be to undo its literary overdetermination as much as it could be to untangle it. In any case, 'unravelling' suggests processing rather than merely recapitulating of London's 'syntax'. A diversity of responses has resulted from this carefully maintained ambiguity.

Peter Barry is paradigmatic of what is perhaps the dominant trend in the critical reception of Iain Sinclair's texts in positioning him amongst a group of London writers who 'are acutely aware of the city as a palimpsest, a document constantly over-written'. Sinclair is found to be one of a group of 'visionary historiographers' who characteristically see urban space as 'constantly deconstructed and reconstructed, always part of a dynamic, in which everything bears the mark of its previous stage of evolution' (*Contemporary British poetry and the city* 165). Sinclair's walks and resulting texts commonly gravitate towards sites of erasure or absence, furnishing, as David James identifies, a 'spectral tropology' that critics have found useful in understanding the intertextualities of Sinclair's writing ('As He Stalks' 155). Hence, if for Barry the city can be read like a text, John Sears can find that Sinclair's representation of the city is substantially *made up* of texts: 'Sinclair's writing is haunted by other texts, saturated by intertextual allusions and citations, and, in some cases, structured around or dependent upon precursor texts for its form and motive force' (172).[1] Alex Murray finds that for Sinclair 'the city is inherently textual', and that 'the textual surface of the city can be read and interpreted' (8–9). Julian Wolfreys's multi-volume *Writing London* argues that London's material spaces are always 'haunted' by textual traces. As Wolfreys describes it, 'understanding the city in terms of text, writing, networks appears to be the only option and one dictated by the city itself'; furthermore, 'the city is a text, inescapably so, performed from traces, marks, citations, narratives legible and illegible' (1: 196, 199). It is the apparent 'revenance' or return of textual traces of place in the works of contemporary writers such as Sinclair (or Sebald, or Carson) that accounts for the prevalent critical vocabulary of spectrality.

However, hauntological methodologies have not gone without critique from prominent detractors. Roger Luckhurst notes that following the publication of Derrida's *Specters of Marx*, the permeation of cultural theory with 'a language of ghosts and the uncanny' has signified 'something of a "spectral turn" in contemporary criticism' (527). Luckhurst's article expresses scepticism that such a spectral turn might be 'anything other than self-referential involution', due to the fact that 'the generalized structure of haunting is symptomatically blind to its generative loci' (528, 530). In other words, Luckhurst argues that

1 Other critics make comparable judgements: Sebastian Groes argues that 'through sheer study, learning and self-immersion in historical texts', Sinclair tries 'to bring back the materiality of history through esoteric acts of narrative' (96); Robert Bond finds a preoccupation with 'all-encompassing textuality' in Sinclair's writing (133); and Christine Ferguson notes a 'metaphysics of textual determinism' there (56). Urszula Terentowicz-Fotyga's more sophisticated argument finds that whilst '[f]or Sinclair … the quest for the real … lies in activating the semiotic codes of urban textuality', his work also makes 'an attempt at grounding the city text in recognizable, or at least recoverable, referents, locating personal inscapes in mappable spaces' (316, 318). Baker's stance is similar: whilst Sinclair 'inherits the tradition of … London as a text', his attempt to 'represent the experience of living in the city' is 'certainly on the side of 'the fragmentary impressions of the walker rather than the totalised vision of the cartographer' ('Maps of the London Underground' n.p.).

those critics who trace spectrality in literary texts are not sufficiently attentive to the genesis and situation of haunting presences. As a result, 'the discourse of spectralized modernity risks investing in the compulsive repetitions of a structure of melancholic entrapment' (535). Luckhurst's critique is thus at odds with Peter Brooker's analysis of an 'aesthetic of recycling' in Sinclair's texts, that serves to 'recover and sustain a valued but disappearing past and to extend or "re-function" this in the present, beyond a circle of self-reference' ('Iain Sinclair' 232–3). Brooker's reading is indebted to Lefebvre's account of the social production of space, the ongoing processes by which societies formulate their own spaces. For Lefebvre, history 'becomes inscribed in space. The past leaves traces; time has its own script. Yet this space is always, now and formerly, a *present* space, given as an immediate whole, complete with its associations and connections in their actuality' (37). 'Making' space, and the influences that inform this process, are at issue in these debates. What is at stake here, then, is the effectiveness with which critics account for and interrogate the selections that contemporary writers make when they invoke a spectral presence, in the form of another writer or text, from a vast array of possible source materials. Processes of selection also necessarily signify elision for which it is necessary to account.

In the context of Sinclair's work, one of the factors influencing processes of selection is the definition of what is permissible as source material. Luckhurst's critique might be extended to suggest that there is an anterior problem when the source material is constituted, by and large, textually. The textual genesis of Sinclair's engagements with London is something that cultural geographers have taken issue with. For Steve Pile, Sinclair's narratives are of interest partly 'because of the hidden connections criss-crossing the city' that they trace, but also partly 'because of the things that he does not see and the things he cannot talk about in his spatial narrative' (117). Pile's terms of description implicate the spectral aspects of Sinclair's texts in the elisions they implicitly make. 'For Sinclair, the city is haunted, and by tracing the stories of the dead he articulates a hidden mood of the city':

> [S]ince only particular things are noticed, those that are have special significance. Sinclair himself is drawn to talk about particular things, to trace only specific histories. … Sinclair is fascinated with those exciting tales of the underworld, of leftist revolutionaries, of artists, writers and film-makers, but not with stories about those mundane trips to the shops with kids in tow, the interminable wait for the bus or the difficulty of using public transport when you're sick and tired. (122–4)

Brian Baker makes a similar point in noting that the 'forgotten London' that Sinclair works to recover is 'only one of very many'. In Sinclair's texts, Baker finds, 'the "real" London (opposed to the official) lies not in the banal, everyday existence of its denizens, but in the visionary "dispossessed" [i.e. forgotten writers and artists]' (*Iain Sinclair* 107). In short, the contention is that despite the fact that they construct an alternative version of London, Sinclair's texts do so by means of recovering lost *texts* and not, crucially, overlooked *practices*. Pile's argument is that quotidian experience ('mundane trips to the shops', 'banal, everyday existence') is always-already excluded in this process of selecting a genealogy. If Sinclair's texts gravitate towards a textualized city in which pre-conceived walks search out the traces of other writers' texts, then the walking practices themselves might be predisposed to confirm what Sinclair's prior research has already suggested. The marginalization of experience in this nexus risks involution. Representation obscuring practice is a concern here.

However, a spectral or exclusively textual reading of Sinclair's texts has not been the only critical response to his work. Other critics have found that representation and experience act as complementary rather than oppositional forces there. Whilst acknowledging that, for Sinclair, walking constitutes 'marching with determination along predetermined routes drawn on maps ... to investigate certain pre-established hypotheses', Robert Sheppard notes that the points of most interest 'are often the accidents, the messages that arise out of experience'. Furthermore, he finds that for Sinclair, 'the only way to establish a working map is empirically, to walk out into the culture ... and gather what the walking reveals' (17–18). David James has made a similar point:

> Sinclair's migratory inquests suggest ... that only by going *in search of* London beneath the everyday rhythms that submerse it, can one begin to research the city's most elusive and illusive inner realms. And by extension, only in committing to searching its social environment at first experientially might the cityscape be re-searched imaginatively. (*Contemporary British Fiction* 71)

If other critics have emphasized the central role played by texts in Sinclair's renderings of space and place, these accounts are invested in redressing that imbalance and paying attention to the everyday practices that also have a distinct role to play. James's account, like Sheppard's, gestures towards the fundamental importance of walking in this experiential processing of the city. These are, it is of note, '*migratory* inquests' – a phrase that holds 'exploration' and 'exploitation' in tension. My own study takes the pedestrian aspect of Sinclair's work as its major focus and aims to assess both its influence on his modes of composition when formulating textual representations of place, and the extent to which he utilizes the space of the text to recuperate ambulatory perspectives and perceptions.

To focus on the role of walking in Sinclair's texts is not, however, to repudiate the intertextual emphases of spectral approaches to his work entirely, but to reframe them. There are many ways to establish dialogues between seemingly disparate representational and non-representational accounts of landscape here. Indeed, the hauntological approach to London writing is not always completely distinct from accounts that emphasize practice and experience. The second and third volumes of Wolfreys's *Writing London* both revise the assertion that 'the city is a text' found in volume one (1: 199). A largely similar argument is rehearsed in volume two where Wolfreys finds that '[i]f the city calls, and calls unceasingly, it does so in ... so many acts of writing generated simultaneously' (2: 7); but in volume three this textual definition of London is expanded:

> [T]he writer of London 'invents' the city, its locations, its memories and histories, its experiences and events. He or she finds them, uncovers their traces and duplicates, reiterates them in the imaginative invention of the particular singularity to which the given narrative attests. This is writing the city as invention. But in that sense of discovery, or finding what is already there, rather than creating from nothing, the other semantic horizon of the phrase makes itself known. (3: 4)

The tension between inventing as discovering and as creating (perhaps exploring and exploiting respectively) informs a lot of what this chapter has to say about processes of articulating place. Sinclair's writing is predicated, I argue, on the discoveries arising out of walking, but also simultaneously engages in the creation of place out of an agenda drawn from textual sources. The link between the two, I suggest, lies in the fact that the 'textual genealogy' that Sinclair constructs, and slots his own texts into, is one that highlights other writers for whom walking is an important constitutive activity. If there is a reliance on texts, it is one that gravitates towards texts that, conversely, stress the importance of practice and process. As a result, where Wolfreys's 'reiteration' is discursive, Sinclair's version of recuperation is also bodily re-enactment. Finally, Sinclair is attentive to the ways that texts selected for extended treatment aim to reproduce the experience of walking in their poetics, and attempts to replicate this in the formal characteristics of his own writing. In short, I contend that representation and practice coalesce in Sinclair's texts in the approaches to walking that they exemplify.

## The Rhythms of Walking

Jeffrey C. Robinson's *The Walk: Notes on a Romantic Image* was one of the earlier texts in a growing field of work renewing scholarly interest in the processes of walking and its implications for literary production. His account opens with an acknowledgement of the seeming tension between walking as an activity retaining a large degree of sensory engagement with place, and simultaneously as a prompt for mental departure, even reverie:

> [T]he walk implies a mixture, or alternation of committed responses and disinterested reflection. ... The walker observes things from a distance, and if the power of the object is in some way too compelling, he by definition detaches himself from it by walking on. Yet the walker is in experience, feels and thinks in his movement through time and space, and is reaching out (or can) to the world in time. To deny either side of the walk is to deny half of the experience. (4)

The tension between exploring and exploiting in Sinclair's formulation registers in different terminology here. 'The walker', Robinson continues, 'often encountering everything afresh is an innocent. But the walker at the same time comes to experience not innocently but full of knowledge and opinions, as a maker of judgements.' The 'immediate pleasures' of walking are, at least in part, produced by the walker's awareness of a 'historical community' of walkers (5). In short, the *immediate* perception of place and landscape that walking allows is actually always *mediated* by a history of walking and, specifically, by prior representations of walking and of place. The tension circulates around the question of whether the gesture towards reverie *precedes* walking, or *proceeds* from it.

Robinson's suggestion that both aspects of walking are rarely thought of together is true of a large proportion of critical accounts, with the majority prioritizing its cerebral over its physical activity.[2] Roger Gilbert is more accurate in his assessment: 'Thinking is … not a mastering of experience but a product of it, circumscribed by the temporal and spatial limitations of the [walking] body itself.' Yet at the same time, '[a] walk is a kind of frame, within which experience takes on a more intensely aesthetic quality' (6, 11). Robin Jarvis proposes that 'pedestrian motion can condition or mediate thought and perception':

> Despite the volitional character of walking, on a long-distance walk the regular, alternating rhythm of right leg, left leg, can induce a hypnotically self-absorbed state … which one typically moves in and out of, and which can take the form of an irregular mental play – the wandering of reverie, or the re-presentation to consciousness of what has been seen and felt – as well as more abstruse forms of introspection and concentrated creative thought. (67–8)

Working from Robinson's commentary on the untenable tendency to separate sensory perception and cerebral computation, Jarvis demonstrates that the sensory engagement attendant to the physical activity of walking, particularly the rhythm it produces, contributes to, or even 'induces', reverie, reflection, and mental departure. Solnit also proposes a model whereby '[t]he rhythm of walking generates the rhythm of thinking': 'Walking itself is the intentional act closest to the unwilled rhythms of the body, to breathing and the beating of the heart. It is bodily labor that produces nothing but thoughts, experiences, arrivals' (5). If these constructions undermine the argument that reverie precedes the walk, then it is equally important to note the (historical) context within which the walk takes place. Solnit also finds that walking on an established route is a kind of reiteration: 'A path is a prior interpretation of the best way to traverse a landscape, and to follow a route is to accept an interpretation, or to stalk your predecessors on it as scholars and trackers and pilgrims do' (68). Whilst it is not clear that retracing a route necessarily denotes accepting an interpretation, it does indicate an engagement with, and intervention in, prior renderings of space. John Wylie's account of the configurations of self and landscape produced by walking, whilst focused primarily on the affects and percepts of sensory engagement, acknowledges that 'attunement' to a history of cultural meanings and sensibilities (including a literary 'romantic inheritance') 'remain[s] the precondition and the *milieu* of contemporary countryside walking in England'. As a result, walking in a landscape 'does not in any straightforward way constitute an "embodied" connection or immersion that is foreign or resistant to the knowledges produced by gazing, contemplating or navigating' ('A single day's walking' 235–40). Embodiment need not occlude cultural knowledges.

[2] Leslie Stephen's essay 'In Praise of Walking' (1902) exemplifies this approach.

Jarvis, Solnit, and Wylie all offer composite accounts of walking that attend to both its physical and cerebral facets, and, crucially, emphasize their processual interaction. In short, a fuller expression of the relationship between walking, thinking, and writing does not need to choose between sensory perception on the one hand, and mental departures and textual influences on the other; but rather involves an acknowledgement of the indivisibility, the necessary interaction, the mutually informed and informing association of the two. An awareness of these twin valences is one of the defining characteristics of Iain Sinclair's textual (re)production of place and landscape in his accounts of walking in London and around its perimeter.

A stretch of the M25 walk on which Sinclair loses his spectacles is, in the corresponding account in *London Orbital*, worked into a comment on these issues. Sinclair decides initially to retrace his steps to retrieve the glasses, but then changes his mind: 'A degree of softness in focus is no problem. It might even be a benefit. Elective Impressionism' (381). The metaphor used to describe what Sinclair sees without his glasses (the alteration to his sensory perception) recalls an artistic style that enacts the same soft focus. The coalescence of experiential perception and inherited knowledges is affirmed further in that Sinclair's blurred vision now approximates the painterly style of Samuel Palmer, whose 'Valley of Vision' artworks the Kent landscape traversed is most associated with, and with whom this section of the text is engaged at length. Furthermore, Sinclair writes that '[t]he rest of our walk is recorded on the same terms: soft shapes, ripe colour, more dream than document'; yet that record is 'truer to the way I felt, the way I *really* saw the road, than all my previous impersonal loggings' (403–4). The representation of these incidents is both dream- or reverie-like and coloured by textual and pictorial influences on the one hand, and centrally concerned with sensory perception and recording the specificity of the encounter on the other. The two aspects are not at odds; rather, they are thoroughly entwined.

A mode of reading walking texts based on an understanding of this interaction steers a course between two potential failings. Anne Wallace uses the term 'falling up' to describe a common critical practice of 'moving immediately from the material terms of the text into the immaterial, the ideological or psychological or whatever, and then staying in abstraction without closely examining the image or plot movement or other material representation which generated that extended meaning' (3). If literary-critical accounts might tend towards this imbalance, the opposite is true of geographical accounts that draw heavily on non-representational theory. In the attempt to focus in on material practices of place and embodied engagement, the context of that event is often overlooked. If narrative is elided, then the account of place arrived at is necessarily denuded. Conversely, an interdisciplinary and combinatory approach recognizes and negotiates both dangers.

This formulation of the roles played by sensory experience and by a prior history of representation in walking is attended by a series of implications for the aesthetics of the walking text. It might be argued that this gesture towards the framing of immediate experience in a history of representation constitutes a way of reconciling the sensory encounter to the already-seen. If literary (and visual) representations of place engage as much with prior representations as they do with the embodied perception of the place, they may also tend towards reproducing earlier versions. It has certainly been argued by a number of geographers that the recourse to representation has tended to flatten and neutralize the disruptive potential of unmediated experience. It is this anxiety that informs, for example, Cresswell's concern that 'landscape', a term he finds to be associated primarily with representation, 'obliterates practice' ('Landscape and the Obliteration of Practice'). The potential for the production of conservative accounts of landscape, reproducing what has gone before in form and content, is a real danger. However, Sinclair's texts do not fall into this kind of involution. The traditions of writing about place and about walking that Sinclair selects for invocation and engagement construct a radical rather than a conservative lineage. In this way, Sinclair is paradigmatic of the contemporary re-emergence of fruitful modes of engaging with place. To evidence this claim is to address the question of exactly which models of walking (and writing about walking) Sinclair is developing.

## Ambulatory Lineages

Critics, guided by certain features of Sinclair's texts, have tended to inaugurate comparative models between the formulations of walking to be found there and the Situationists' concept of psychogeography, itself developed out of a French tradition of urban walking falling under the umbrella concept of *flânerie*.[3] I contest that a fuller sense of Sinclair's relationship to antecedent walking practices only emerges when this comparison is set within a longer history of pedestrianism. Whilst his texts undoubtedly engage with this French tradition, they also draw heavily on an English Early Modern and Romantic peripatetic legacy. These two genealogies impact upon one another in Sinclair's work, and a new hybrid model of ambulatory engagement with place emerges there. An examination of the methodological statements embedded in Sinclair's topographical non-fiction reveals the extent to which he is concerned to move away from the objective account of place aspired to in the 'botanizing on the asphalt' of the Benjaminian

[3] Examples of this privileging of one model of walking above others include Brian Baker's assertion that 'Sinclair's use of psychogeography is derived from the writings of the Situationist International' ('Maps of the London Underground' n.p.), and Robert Sheppard's likening of Sinclair's walking to 'the modality of the [Situationist's] *dérive*' (85). Baker does note, however, that in *London Orbital*, Sinclair has gone some way to repudiate his former use of the *flâneur*. Merlin Coverley's *Psychogeography* has provided a longer history of this kind of walking and has thus gone some way towards resisting exclusive Situationist ownership of the term. In the same vein, Karen Newman has noted that '[t]hough the first uses of the word *flâneur* do in fact date from the early part of the nineteenth century, *flânerie* and the verb *flâner* … date from the early seventeenth century' (61–2).

*flâneur* (*Charles Baudelaire* 36), and in the 'precise laws and specific effects of the geographic milieu' in Guy Debord's brand of psychogeography ('Introduction to a Critique of Urban Geography' 59). In their stead, a key passage of *London Orbital* explains the circular walk as 'an hysterical fugue' and replaces the detached *flâneur* with the subjectivity of what Sinclair calls the *fugueur* – one for whom the walk is a 'once-a-month episode of transient mental illness' (146–7). This realignment signifies a shift towards the altered perspectives found in the walks of figures such as Thomas De Quincey and John Clare, whose 'Journey out of Essex' Sinclair subsequently retraces in *Edge of the Orison*. It is, I argue, from this avowedly idiosyncratic basis that Sinclair's own walking texts emerge.

Sinclair has called De Quincey the father of the kind of walking that he himself undertakes, and often gravitates towards writers of the Romantic era (William Blake and John Clare consistently) and the Early Modern period (John Bunyan's *Pilgrim's Progress* is, for him, 'the ultimate English walking book') (qtd. in Nicholson 95–6). The canon of walking writers he selects share political or aesthetic attributes rather than emanating from a single historical period. A text like *London Orbital* does emphasize Blake, Clare, and other eighteenth- and nineteenth-century writers and artists such as Samuel Palmer, but it also establishes links between their radical modes of walking and a dissenting pedestrian tradition through studies of, for example, Wat Tyler in the fourteenth and the Diggers in the seventeenth century. Whilst a set of meanings that come to prevail in the Romantic period are subscribed to by Sinclair, certain of the connotations that fall away at this period are also resurrected in his interpretation and use of walking practices.

Wallace describes 'a genealogy of writings about walking, with a significant shift from negative representations to positive [taking place] in England in the first few decades of the nineteenth century in congruence with ongoing material changes that caused or permitted changes in the material and ideological shape of walking' (9–10). The most influential of the material changes that Wallace goes on to detail is the increased availability of affordable coach (and later rail) travel, and the effect this has of removing the association of walking with necessity. Prior to this shift, 'the subversive connotations of walking' included poverty, homelessness, and criminality (54). Carl Moritz's 1782 journal of a pedestrian journey through England is the resource most often turned to by scholars of Romantic walking to demonstrate the 'subversive' associations prior to the cultural changes thought to have taken place in the 1790s. Moritz described how 'any person undertaking so long a journey on foot, is sure to be looked upon, and considered as either a beggar, or a vagabond, or some necessitous wretch, which is a character not much more popular than that of a rogue' (173).

Other researchers have contested Wallace's dating of these changes: Donna Landry, for example, attests to 'a social and semiotic shift' taking place across the whole of the eighteenth century, whilst early modern scholars recuperate still earlier examples of aesthetic registers of walking practices (22–3). Jarvis, who challenges Wallace's view that the break is complete in the 1790s, finds that the Romantics inherit and play off connotations of 'poverty, unrespectability, and possible criminal intent' (22–3). Later in the nineteenth century, the dominant associations attached to walking shift towards its wholesomeness – Solnit notes that '[f]ew of the canonical essayists [of this period] can resist telling us that we should be walking because it is good for us' (121). What is significant for this study is that Sinclair, like a number of Romantic writers, also plays off a set of subversive associations, whilst downplaying or even completely overlooking the improving, cultivated modes of walking that come to the fore later. In *London Orbital*, Sinclair explicitly distances his walking practice from an 'Edwardian sense of excursion, pleasure, time out' (35). In short, the textual genealogy of walking that his texts construct incorporates a strong Romantic inflection because this period offers some of the most accented versions of subversive walking that Sinclair sees himself as buying into.

Sinclair has described his writing as 'on-road and off-message' ('At the Edge of the Edge' n.p.). This pairing draws attention to the central role of walking practices in the production of an 'off-message' rhetoric and an alternative perspective on the landscape encountered, as well as the society and culture producing it. This connection is emphasized throughout Sinclair's walking texts. In *London Orbital*, for example, the practice of walking around the M25, where the car and vehicular travel are normative, is integral to the reading of that landscape. The text searches out the hidden or overlooked places, and brings to light that which official narratives elide. A technique deployed repeatedly in Sinclair's writing to describe his methodologies is the recording of another writer or artist's statements about their practices that might just as easily illustrate his own. In *Hackney, That Rose-Red Empire* (2009), Sinclair quotes from Patrick Wright's explanation of his mode of writing about place: 'I take oblique approaches, not out of perversity … but because they enable me to cast unexpectedly revealing light on apparently familiar realities, and thereby free my subject matter from the grip of ideological orthodoxy' (210). Wright's statement, and Sinclair's use of it, is informed by Victor Shklovsky's concept of defamiliarization, wherein '[a]rt removes objects from the automatism of perception'. In making an object unfamiliar, Shklovsky finds, art undermines habitual perception such that 'one may recover the sensation of life; it exists to make one feel things, to make the stone *stony*. The purpose of art is to impart the sensation of things as they are perceived and not as they are known' (12–13). These terms of description interestingly inflect the distinctions between discovering and creating the city discussed above, and the interaction between representation and experience that underpins this chapter and this book at large. Here, the defamiliarizing techniques of literary representation offer not a departure from practice and experience, but a restoration of awareness and focus in perception. For Sinclair, it is walking that signifies an 'oblique approach' and produces a defamiliarizing effect. The regular, punctuating references to earlier writers whose walking produces similarly alternative accounts of place, align Sinclair with a tradition of subversive pedestrianism.

It is in these defamiliarizing capabilities attached to walking that comparisons between Sinclair's walking and (Situationist) psychogeography have been drawn. *Lights Out for the Territory* is explicit in its engagement with *flânerie* and psychogeography, but is also at pains to point out key differences. The *flâneur* is mythologized in the writings of Charles Baudelaire, and subsequently of Walter Benjamin, as a privileged and disinterested urban wanderer amidst the crowd. 'Botanizing on the asphalt', Walter Benjamin's aforementioned famous phrase describing his activities (and it was, almost exclusively, his), is glossed by Mike Crang as 'subject[ing] urban life to the detached curiosity and categorizations normally reserved for the natural world' (54). Thus, in Benjamin's account, the *flâneur* 'claimed that he could divide the Parisian public according to its various strata as easily as a geologist distinguishes the layers in rocks' (*Charles Baudelaire* 39).[4] The same gesture to a walking method producing an objective account of place and a new kind of urban sociology is carried over into the post-war psychogeography of Guy Debord and the Situationists, shortly after Benjamin's rediscovery of Baudelaire and the Surrealists' reinvention of *flânerie*. Debord's oft-quoted definition of psychogeography as formulating 'the materialist perspective of the conditioning of life and thought by objective nature' in urban space is comparable to the classificatory and codifying aspects of the *flâneur*'s activity in 'botanizing on the asphalt' ('Introduction to the Critique of Urban Geography' 59). Just as the latter metaphor draws on scientific disciplines, Debord's psychogeography is akin to the recognition of 'determinant action of general natural forces' in geography: 'Psychogeography will aim to study the precise laws and specific effects of the geographic milieu, consciously planned or not, acting directly on the affective comportment of individuals.' Following from this, the adjective 'psychogeographic' is found to describe 'the data ascertained through this type of investigation' (59). The ambiance of a locale 'can be revealed by thorough analysis and turned to account' (61). Whilst the terminology of 'precise laws and specific effects', of 'data' and 'investigations', is far more pronounced in the latter manifestation, continuity between the *flâneur* and Situationist psychogeographer both aspire to an objective account of a place's resonances. This aim is one that Sinclair's texts eschew in their explicit construction of subjectivity.

If Debord's elaboration of the *dérive* (literally translated, 'drift') is echoed in Sinclair's description of his own methodology in *Lights Out for the Territory*, a rebuff to impartiality is also included:

> Drifting purposefully is the recommended mode, tramping asphalted earth in alert reverie, allowing the fiction of an underlying pattern to reveal itself. To the no-bullshit materialist this sounds suspiciously like *fin-de-siècle* decadence, a poetic of entropy – but the born-again *flâneur* is a stubborn creature, less interested in texture and fabric, eavesdropping on philosophical conversation pieces, than in noticing *everything*. (4)

Sinclair's terms of description suture together the three figures of environment, walker-writer, and text. The city possesses an 'underlying pattern' that, when

[4] Elsewhere, Benjamin also acknowledges the complications attendant to the stance taken up by the *flâneur* in relation to the crowd: 'He becomes their accomplice even as he dissociates himself from them' (*Illuminations* 168).

described as a 'fiction', is linked to its reproduction (with embellishments) in the text produced by the writer. The activity by which this interaction of environment and subject in the production of text occurs is 'purposeful drifting' – as Sinclair goes on to note, '[w]alking, moving across a retreating townscape, stitches it all together' (*Lights Out* 4). Here, stitching carries the same resonances for Sinclair that weaving does for Carson. The ambiguous adjective 'retreating' in this phrase returns this discussion to the unresolved tension between exploring and exploiting as modes of engagement with which this chapter opened. A 'retreating townscape' might be one whose particularity recedes before the encroachments of Sinclair's fictionalizing tendencies of reportage (his rhetorical 'exploitation' of the city). Alternately, the phrase might indicate the slow unfolding of perspective that travel at a walking pace dictates. This second reading emphasizes the characteristics of sensory perception that are attendant to walking and attuned to exploration. Whilst Sinclair's refusal to explicitly choose one or the other mode leaves his writing open to critique, the careful ambivalence of this phrase is indicative of a combinatory approach.

Sinclair's application of the adjective 'purposeful' to the activity of 'drifting' suggests his distance from the nineteenth-century *flâneur*, who is parodied as an eavesdropper and decadent figure in this passage. The idea of 'drifting' is retained for its potential as an oppositional activity to capitalistic modes of living: 'Time on these excursions should be allowed to unravel at its own speed, that's the whole point of the exercise. To shift away from the culture of consumption into a meandering stream' (*Lights Out* 7).[5] Elsewhere in *Lights Out*, Sinclair goes further in his repudiation of the role of *flâneur*: 'The concept of "strolling", aimless urban wandering, the *flâneur*, had been superseded [*sic*]. We had moved into the age of the stalker; journeys made with intent – sharp eyed and unsponsored … This was walking with a thesis' (75). For Chris Jenks, 'the *flâneur* possesses a power, it walks at will, freely and seemingly without purpose, but simultaneously with an inquisitive wonder and an infinite capacity to absorb the activities of the collective – often formulated as "the crowd"' (146). The *flâneur*'s relation to the city is one of absorption of its energies and its inhabitants' activities; he is the vessel for the recording of the city as it is. The *flâneur* is passively receptive in contrast to Sinclair's active rendering of the city. His interaction with the urban environment is one of collaboration, wherein 'recording and retrieving the messages' already inscribed in the palimpsest of the city is a component, but this act of 'alternate reading' is measured against one of Sinclair's own writing activity: 'to vandalise dormant energies by an act of ambulant signmaking' (*Lights Out* 1). It is at this juncture that criticisms of Sinclair's method find most purchase: 'walking with a thesis' seems to overstep the balance identified above, side too heavily with exploitation, and marginalize the exploratory aspect of walking. Whilst there is justification for this kind of disapproval, there is also a possible rebuttal.

---

5 This set of meanings does indicate some continuity with the Benjaminian *flâneur*, whose slow, drifting pace was a 'protest against … industriousness' (*Charles Baudelaire* 54).

The reading of space that Sinclair engages in is one in which selectivity, arrangement, and orchestration are active components. He describes his walks variously as 'phantom biopsy' or 'visionary exploration' (*Lights Out* 4–5). These composite phrases combine terms from scientific or geographical discourse with a vocabulary of mysticism. As such, they signify the dual mode of apprehending urban space: objective documentation is supplemented by creative rendering. Whilst *flânerie* and continental constructions of psychogeography undoubtedly have a place in a discussion of Sinclair's ambulatory practices, the visionary aspects of walking, and their production of 'alert reverie' denote a modulation towards an English tradition most strongly realized in peripatetic Romanticism.

Just as the walker's perspective that Sinclair installs is provisional, so its definition shifts over the run of his texts. If *Lights Out for the Territory* establishes a method, *London Orbital* and *Edge of the Orison* extend and refine it. It is on the Shenley section of the M25 walk that a new terminology emerges, extending the subjective theme into what Sinclair calls 'mad walking'. The 'orbital pilgrimage' could be explained as 'an hysterical fugue – attended by … minor epileptic seizures (electrical storms in the consciousness)' (*London Orbital* 146–7). Sinclair's use of the term 'fugue' draws on its psychiatric definition as a neurotic's 'flight from one's own identity, often involving travel to some unconsciously desired locality'. As such, the M25 circuit walk becomes a stepping outside of everyday experience and consciousness to read different narratives of place. The fugue metaphor thus combines the anti-psychiatry of R.D. Laing – an influence on Sinclair since the early phase of his writing career – with walking processes to reaffirm the defamiliarization identified as one of Sinclair's key aims.[6] On this walk, the objective *flâneur* is exchanged for an alternate mental state with different attentions and perceptiveness:

> *Fugueur* was the right job description for our walk, our once-a-month episodes of transient mental illness. Madness as a voyage. The increasing lunacy of city life … forced us to take to the road. The joy of these days out lay in the heightened experience of present tense actuality, the way that we bypassed, for a brief space of time, the illusionism of the spin doctors, media operators and salaried liars. (*London Orbital* 146–7)

The resulting account of place now only proposes an alternate reading (from amongst myriad potential meanings). Furthermore, 'mad walking' emanates from the landscape traversed: walking around the M25, Sinclair finds that 'asylums haunt the motorway like abandoned forts' (163). The mode of experience that Sinclair selects is thus in harmony with the place itself.

In positioning himself as a '*fugueur*', Sinclair is tapping into a lineage of such figures to define his own approach to recording London. His notion of the 'lunacy of city life' forcing the *fugueur* to 'take to the road' as a way of 'bypassing'

---

6 Laing suggested that schizophrenia and other 'mad' perspectives offered a redemptive alternative to normative and indoctrinated 'sane' positions (cf. *The Divided Self*).

everyday pressures is indeed one with a rich literary pedigree. The idea that the *fugueur* 'simply walk[s] out of his quotidian life' (146), and enters into an alternate mental state whilst walking, is one that might be related to Debord's assertion that the *dérive* 'contrasts on all points with the classical ideas of the journey and the stroll', that is, that it signifies a departure from habitual uses of space ('Theory of the *Dérive*' 78). In Sinclair's restatement – 'the fugue is both drift and fracture' – the extent of the mental departure is emphasized (147). Here, the legacy of a number of Romantic prose writers is also felt. Charles Lamb's 1825 essay 'The Superannuated Man' makes a resonant association between quotidian employment and 'captivity'. 'The irksome confinement of an office' with 'no prospect of emancipation', likened to 'the condition of a prisoner in the old Bastille', is internalized by the office clerk: 'I had grown to my desk, as it were; and the wood had entered my soul.' The clerk's release is in Sunday walking excursions – in Lamb's terms, his 'poor Sabbath recreations' when, in a phrase pertinent to this discussion, 'I walk ... just when the fit seizes me' (186–8). Indeed, Lamb's biography reveals him to be something of a model for Sinclair's *fugueur*: Lamb spent six weeks in an asylum in Hoxton in 1795 due to poor mental health, and a year later his sister Mary murdered their mother in a fit of mania (after which the jury at her trial recorded a verdict of insanity). In *London Orbital*, Sinclair notes that 'Lamb has been heritaged as one of the treasures of Enfield' (95). Remembering another less-known and less-celebrated version of his persona is thus especially important.

Another Romantic prose writer, William Hazlitt, in 'On Going a Journey' (1822), finds that '[w]ith a change of place we change our ideas' and that '[w]e go a journey chiefly to be free of all impediments and of all inconveniences' (147, 141). Stepping outside of everyday life on a Sunday excursion registers in Sinclair's text as a regular, that is, 'once-a-month', bout of 'transient mental illness' (147). The precedent for considering this walk as an altered mental state and identity is also found in Romantic writings. Hazlitt considers just such loosening of the bounds of identity on a peripatetic journey: 'Oh! it is great to shake off the trammels of the world and of public opinion – to lose our importunate, tormenting, everlasting personal identity in the elements of nature, and become the creature of the moment, clear of all ties' (146). Becoming 'the creature of the moment' echoes in Sinclair's 'heightened experience of present tense actuality'. The textual traces of antecedent walkers present in Sinclair's embedded methodological statements – such as this – work to foreground the degree to which sensory perception and a sense of the walk as an individuated event are retained even where a pre-conceived agenda initially seems to marginalize such concerns. The intertextual departures that at first appear to lead away from practice are in fact included largely to emphasize the processual nature of walking. This is the reciprocal play of representation and experience in Sinclair's walking texts.

Sinclair declares that 'superstition persists: the walk around London's orbital motorway is *personal*'. Instead of any 'objective method', he opts for 'mystical correspondences' (*London Orbital* 255). Just as the walking procedures are subjective, so is the text that they produce. In *Edge of the Orison*, different participants bring their own differing agendas to the walk 'under a single flag of convenience' (53). This results in complications: '[t]he problem for our pedestrian trio … is that there are no sidekicks: or, we're all sidekicks but won't admit it' (133). Partisanship is further foregrounded by the attention conferred upon Sinclair's co-walkers' very different reports of the same journeys. In *London Orbital*, he notes that 'Kevin Jackson's account of the day's walk', an intertext for his own version, 'is very good on our induction into Staines café society'. Furthermore, this allows for a comparison between different acts of recording and modes of composition: 'Kevin, I realise, is taking everything down in a neat notebook', whereas Sinclair prefers the 'selective memories' that arise from writing up notes '*after* the event' (249). The contrast only serves to throw into relief the artistry of Sinclair's narrative.

Yet this tactic too has a precedent: in a reworking of Wordsworthian recollection-in-tranquillity, Hazlitt declares himself to be in favour of 'the synthetical method on a journey, in preference to the analytical. I am content to lay in a stock of ideas then, and to examine and anatomise them afterwards' (143). This, then, is not the 'botanizing *on* the asphalt' of the *flâneur*, but a process of anatomizing that takes place retrospectively – the partiality and selectivity of memory are introduced at the very heart of the process of recording, producing a text that recalls the 'bravado style' associated, in an earlier era, with the 'Cockney School' Romantics (Jarvis 213). Sinclair's writing operates with an implied relationship between representation and experience: an idiosyncratic style of reportage underscores a subjective rendering of place produced out of a complex of sensory perception and literary reverie.

## *London Orbital*

At several dispersed points in its narrative, *London Orbital* incorporates attempts to define the purpose of the walk that it reports on. The M25 walk is, for Sinclair, both tourism and exorcism. If the former is related to the Sunday and holiday excursion of Cockney School Romantics, the latter plays off the 'purposeful drifting' of *Lights Out*:

> We're walking tourists. We pass through landscapes on which we have no claim. We spend money in pubs. We visit the obligatory sights: churches, parks, bunkers, villages with literary or painterly associations. We take photographs. But alongside the convivial agenda, is a ritual purpose: to exorcise the unthinking malignancy of the [Millennium] Dome, to celebrate the sprawl of London. (*London Orbital* 412)

Both aspects of the walk bear Romantic registers. The 'convivial agenda', the picaresque-like passages of tourism, select an itinerary that traces places 'with literary or painterly associations', a large proportion of which are drawn from the early nineteenth century. At Shoreham, where the painter Samuel Palmer lived and worked between 1826 and 1835, Sinclair and his co-walkers are 'misaligned Ancients, retro-Romantics' (446), 'a self-conscious restatement of Samuel Palmer's gang, the Shoreham Ancients; city folk up for a ramble' (410). This group dynamic recalls what Jarvis calls the 'good-humoured, sociable pedestrianism' of Leigh Hunt, in which the earlier 'Romantic ideology of freedom-through-walking' has been replaced with 'a lighter recreational-touristic creed' (211).

The 'ritual purpose' of the walking project detailed here inaugurates a set of equivalences with earlier spatial practices: in particular, walking the perimeter in this manner recalls the parish custom of 'beating the bounds', familiar since the medieval period. The same analogy can be found elsewhere in Sinclair's work: in *Hackney, That Rose-Red Empire*, for example, a walk tracing the borough's border is described explicitly as 'beat[ing] the bounds' (441). This practice is that of rogation processions of the medieval and early modern periods. Walsham writes that '[y]early perambulation of the community's uttermost limits and landmarks helped mentally to rechart the map that divided neighbouring communities', the procession was 'a key element in the regular reanimation of local identity and collective cartographic memory' (251–2). Sinclair's aim, in invoking these practices, is to draw out a similar process of recuperation in his own walk. The 'recharting' that occurs here – collective in being shared with co-walkers – aims to unearth the forgotten histories of London's edgelands that are missing from the developers' maps.

As a result, the two agendas are not easily reconciled to one another. 'Passing through' places 'on which we have no claim', in the manner of tourists, is at odds with ritual practices of circumnavigation that incontrovertibly stake a claim on a known area. In contradistinction to the human and imagined values of these landscapes, the Millennium Dome stands, in Romantic fashion, as a sublime object – fearful yet fascinating – when Sinclair's treatment of it is so rhetorically hyperbolic. It is resistance to the adverse forces acting on these places that signifies the walk's 'ritual purpose'. Considering the term 'ritual', Victor Turner finds that its symbols are 'not only a set of cognitive classifications', but also 'a set of evocative devices for rousing, channelling, and domesticating powerful emotions' (*The Ritual Process* 42–3). Here, then, the explicit invocation of a vocabulary of ritual raises the stakes. The walk is a ritual 'performance process' in which, as Turner remarks elsewhere, 'what is normally sealed up, inaccessible to everyday observation and reasoning, in the depth of sociocultural life, is drawn forth' (*From Ritual to Theatre* 13). It is exactly this 'drawing forth' of hidden histories that Sinclair aims at.

Like the touristic aspect of the walk, its 'ritual purpose' – the excision of malign influences symbolized in the imposition of the Millennium Dome – is also explicitly informed by one Romantic figure in particular: John Clare. Moving counter-clockwise from Waltham Abbey, the M25 circuit is choreographed to end at Epping Forest. Virtually the last visited site on the circumnavigation, High Beach asylum was the point of departure for Clare's 'Journey Out of Essex' in 1841:

> We calculated that this journey, which we were determined to repeat, was around 120 miles. Or the distance of the M25 if it were stretched out into a straight line.

> Fugue as exorcism: Clare's walk successfully performed the ritual we were toying with. He'd been in the forest long enough to understand the peculiarity of its status as a memorial to a featureless and unreachable past, a living stormbreak at the limit of urban projection. (534)

Fugue and exorcism are linked: the demons that all Sinclair's walks aim to exorcise – the heritage industry and the co-option of history in the service of shoring up and disguising top-down government management of space – are the contemporary manifestations of similar processes taking place in the early nineteenth century, expressed here in the memorial function of the Epping Forest as symbolizing an 'unreachable past' to which Clare is attached and from which he has been severed.[7] That is, Clare's documentation of, and resistance to, the processes of centrally managed enclosure that were irrevocably changing the landscape that he knew, are bound up inseparably with his walking the land. As such, Clare's walking and writing practices provide the model for Sinclair's own resistance to contemporary landscape alterations. The fugue-state of the walk is the means by which these processes are both recognized and represented. In suggesting that Clare 'did what any sane man would have done, he took off on his epic journey', Sinclair inverts the dialectic of sanity and madness (534). The ultimate outcome is that the supposedly mad state of the *fugueur* might actually constitute the most accurate way of perceiving the landscape through which the walk progresses: it is 'the illusionism of the spin doctors, media operators and salaried liars', which Sinclair and his accomplices flee, that bears the mark of insanity (147). The play between the objective vision of the *flâneur* and psychogeographer and the subjective account that Sinclair's text orchestrates is turned on its head.

Clare is the writer most closely associated with the experience of and resistance to enclosure. Enclosure, the central management of and legislation for the increased productivity and privatization of space, is, for Sinclair, a motif that echoes in contemporary trajectories of land use. Wallace notes that 'enclosure revealed walking as an instrument of reappropriation of common lands and perspectives', and that walking through enclosed land possessed a 'material capacity to appropriate paths for common use' (67, 118). Sinclair's retracing of Clare's 'Journey Out of Essex' in *Edge of the Orison* similarly challenges accepted models of land ownership. The patterning of enclosures across historical periods is a repeated refrain in Sinclair's texts. The closing phases of *Hackney, That Rose-Red Empire* detail the 'Olympic enclosures' enacted on that borough of East London, and find common cause between historical and contemporary resistance. Events of the two different periods are described in markedly similar terms: 'They tried to fence off Hackney Downs at the time of the enclosures'; 'the Hackney Wick edge-lands were enclosed' for the building of the Olympic site, and '[o]vergrown paths and ancient rights of way were blocked by boards and fences' (550, 561).

---

7 The history of the Epping Forest is actually not as clear cut as Sinclair makes it: the Elizabethan forest was not common land before enclosure as he infers, but was in fact a royally preserved hunting ground.

The same antipathy between enclosure and rights of way that Clare worked off is present throughout Sinclair's texts.

Considering Sinclair's opposition to the Millennium Dome and the role it plays in *London Orbital*, Baker finds that this building 'erases the social and historical particularities of Bugsby's Marshes [the suppressed name for the site on which it was built] and implants a simulacrum of lived space' (*Iain Sinclair* 85).[8] 'The Millennium Dome and the Asylum', Baker continues, 'are crucial sites in *London Orbital*, signalling the erasure of history and community in contemporary Britain and its replacement with a series of spectacles and simulacra.' In this process, 'cultural memories and practices of everyday life are overwritten and countermanded by contemporary configurations of power, manifested in microcosmic form in the Dome' (152). If the Dome signifies forgetting, the *London Orbital* walk enacts a ritual act of remembering. It is with this rationale in mind that the text culminates in the arrival at High Beach Asylum and takes in so many other forgotten asylums (many now converted into housing estates) in its circumnavigation of London. With this tactic of remembering, Sinclair aims to recuperate a sense of place derived from textual sources ('cultural memories') and sensory perceptions ('practices of everyday life'). The two often interact and play off one another.

Sinclair's circumnavigation of the capital 'within the acoustic footprint of the M25', is a project that works against passively accepting the non-place status of London's hinterlands, but instead aims to recuperate experience of their textures and remember their histories.[9] Sinclair is asked about the motivation for the walk by Bill Griffiths, a co-walker on an early stretch of the M25 circuit: 'Why *counterclockwise* he wants to know. ... The answer to Bill's question has something to do with the Italianate towers, the only surviving markers of hospitals and factories. Our walk is a way of winding the clock back' (69). The orbital walk reads the traces and fragments ('surviving markers') of former guises of place to reconstruct histories ('wind the clock back'). If real estate developers are refitting disused asylums like Claybury and creating a 'stylish no-place that is everyplace' with '[n]o attachment to the local, [but] an easy commute to the centre', then Sinclair is at pains to invoke these sites' former histories that are being forgotten in the process (168).

Crucially, Sinclair's project is reliant on walking as an alternative practice of space opposed to the anaesthetized car-bound perspectives of those merely passing through rather than slowing down or stopping to look and investigate. Sinclair walks, he notes in *Edge of the Orison*, 'to appreciate everything [that] we fail to notice when we pass through here, foot down, in our cars' (18). Sinclair is

---

[8] Sinclair's oeuvre might be conceived in these terms, from the resistance to the Thatcher-era restructuring of London's Docklands in *Downriver* (1991), to Sinclair's hostility to the Olympic infrastructure in Hackney and Stratford in *Ghost Milk: Calling Time on the Grand Project* (2011).

[9] On the idea of 'non-places', see Marc Augé, *Non-Places*. For important formulations of hinterlands, see Williams, *The Country and the City*, and Cronon, *Nature's Metropolis*.

constantly attentive to the way in which 'status changes' between vehicular and pedestrian perspectives (*London Orbital* 140). *London Orbital* testifies numerous times to the way drivers 'ignore landscape. It happens, but it is as featureless as television; no better, no worse. Narcoleptic resignation, postponed pleasure' (323). 'Follow the signs for LONDON ORBITAL in your car', Sinclair writes, 'and consciousness takes a drive' (77). The contrast is clear: for drivers, journeys are 'measured by the hour', and distances between locations approximate '[u]nits of nuisance between pit stops' in 'a geography defined by junction numbers on the M25' (392); whereas Sinclair and his co-walkers 'have followed a preordained narrative and not skipped from first sentence to conclusion' (172). In line with the unresolved tension between exploration and exploitation discussed above, it is left unclear in this passage if it is the walkers or the environment itself that has 'preordained' the narrative read there, but it is certain that seeking and recording fuller narratives is in opposition to abridged 'non-place' designations of the interstitial zone between the capital and the country.

Yet there is a danger attendant to such a project; and one that, in *London Orbital* if not in earlier texts, Sinclair is aware of. He observes that '[a]ny convenient fable can be pressed into service to lend narrative to a resonant location', and so add value to a place's marketability (*London Orbital* 529). '[H]istory [is being] heritaged for anyone with the price of a ticket. … History is an extra zero on your property prices' (150–51). History and heritage are in danger of coalescing. A project that rediscovers histories must therefore find ways to separate itself from the discourse of heritage that co-opts historical narratives into promotional material for real estate. Sinclair records the 'heritaging' process in action at Shenley, where the hospital, 'active until a couple of months before our arrival, has vanished, replaced by a housing development, the bright new units of a Crest Homes estate. The back story of the asylum has been totally erased' (150). The erasure of an inconvenient history runs concurrently with the assertion of heritage:

> History is being revised on a daily basis, through the northern quadrant of the motorway, by copywriters employed by the developers. 'The historic village of Shenley combines excellent local interest with outstanding travel convenience.' Much is made of the 'pleasant undulating countryside' and the 'fine views northward over the historic city of St Albans'. (151)

The revision of history creates a pastoral image of the countryside and a 'historic' city viewed from afar and presumably from an elevated prospect. This framing of the landscape is the very opposite of the walker's engaged and involved perspective.

The same process is underway at Painshill Landscape Park, visited on a later stretch of the journey:

> Painshill (supported by the National Heritage Memorial Fund, English Heritage, the Countryside Commission and Surrey County Council) is overendowed real estate. Like so much else that we've encountered, it is dedicated to a way of customising the past as a way of making us feel good about ourselves: we come from somewhere, we have a lineage. That which is worth preserving has

> been preserved. We meditate, by walking specified and guided routes, on the lessons history can teach us. (318)

In this passage Sinclair emphasizes the role played by governmental authorities and quangos in the institutionalization of heritage. History is 'customised' in the service of the interests of the present – in this case, the development of London's commuter belt. It is also significant that walking 'specified and guided routes' has a role to play in affirming this process. 'Meditating' on 'lessons' handed down actually constitutes a type of mediation similar, for Sinclair, to the 'filtering' that car travel enacts. Walking is a contested practice. In contrast to this 'guided' walking, and the occlusion of counter-narratives of place that it signifies, Sinclair's own walk traverses areas designed for and dominated by the car: the M25 borderlands of London are described as 'a landscape that challenges or defies walkers' (239). The seemingly forbidden practice of off-route walking provides a fuller account of place. The same opposition between two geographies and two kinds of walking carries over into *Edge of the Orison*. When a 'sanctioned path rubs against the motorway, before twisting back among fields and ponds', Sinclair finds that '[b]eing allowed, even encouraged, to move in a particular direction kills the desire' (12). Instead, he chooses to follow 'a drovers' path between rough hedges' (20). This selection foregoes the pastoral scenes of 'fields and ponds' and opts instead for a different aesthetic of 'rough hedges'. In retracing a drovers' route, it also reinscribes traces of labour erased from the landscape. Sinclair's intervention is twofold. He engages with the pastoral or bucolic rhetoric that invariably accompanies the marketing of non-places; and he traces alternative (but equally subjective) histories – often emphasizing labour and dissent – to those that have been 'heritaged'.

Sinclair's aesthetic engagement with London's edgelands is akin to a process described by Barry as 'double visioning' where traces of a place's history remain visible even where it has been much altered. For Barry, this might take the form of 'a verbal trace in an apparently inappropriate street name', or in a physical residue in the terrain's layout, such as 'an odd bend in a road whose building line follows the course of a now-vanished stream' (*Contemporary British poetry and the city* 46). Thus, '"double visioning" simultaneously perceives both the built-up present and the inscriptions upon it of a pre-urban past'. It is 'the attainment of a multi-layered chronological perspective which typically superimposes one historical period upon another, so that the viewed entity becomes radically trans-historical' (46). In similar terms, Sinclair describes his own work as creating 'topographical records of simultaneity' (*Edge of the Orison* 6). Barry's findings, predicated on the city, are also resonant for suburban locations, such as the M25 ring, where city and country are in processual dialogue rather than urbanization being a completed and closed process. Sinclair's texts follow up both the 'verbal traces' and residual topographical features that Barry identifies.

The first walked section of *London Orbital* – from the inner zone of the metropolis to the margin – is one that plays out this process. It traces older routes up the Lea valley and its river rather than more recently imposed roadways, and the first point on the peripheral loop that follows is Waltham Abbey – a now-decayed site of power and the symbolic representative of a mostly vanished order of society. The Abbey is carefully selected as the circular walk's point of departure for the histories it resonates. It signifies a remaining trace of a pre-Reformation moment before its dissolution in 1540, thus implicitly signifying a submerged recusant narrative. Waltham Abbey is also, as *London Orbital* makes clear, the site of the grave of King Harold – that is, Harold Godwinson, the last Anglo-Saxon king defeated at the battle of Hastings, and therefore a reference point for a history of resistance to the conquest and the Norman yoke. It is also worth considering the significance of the walk beginning and ending at a sacred place – that is, the Abbey – in light of the terminology which casts the journey's purpose as one of 'ritual', and as an 'orbital pilgrimage' (146). This location is certainly akin to the numinous places that a rogation procession would halt at for reflection.

The retail parks and research institutes that constitute the walk's subsequent destinations thus sit alongside the remnants of an older infrastructure that mark the circuit's beginning and end. In Sinclair's hands, trans-historical simultaneity is not so much superimposed as entwined. For Sinclair, the old is often beside rather than beneath the new: 'a network of green lanes can sidle *alongside* the densest clots of population' (138, my italics). Non-place and pastoral are mingled in *London Orbital*, the reality of the former and the aesthetic gesture towards the latter (in the form of heritage), are both undercut by their mutual presence and entanglement. From the perspective of Sinclair's double visioning, these places are neither devoid of history, nor do they conform to easily marketable or hegemonic narratives.

The stretch of the M25 walk between Waltham Abbey and Potters Bar provides an example of Sinclair's 'liminal landscape' aesthetic (138). Cattlegate Road (note the pastoral allusion) 'shivers with ambiguity' (135):

> A constant stream, both ways, clusters of five or six cars, nose to tail in barely controlled frustration: metal projectiles time-warped on to a drovers' track, Cattlegate Road. In the intervals between the blam-blam-blam, Renchi [Sinclair's co-walker] hears a woodpecker in the twilight woods. (137)

The speed of vehicles described as 'metal projectiles' contrasts the slow progress made by herded cattle on an unpaved 'track'. However, the pastoral and the modern also entwine in this passage. The cars form a 'constant stream', with the primary sense of their continual passing, but also suggesting road as a waterway. This metaphor is present at other junctures of *London Orbital* and *Edge of the Orison*: 'Soft estate walking [i.e. at the motorway's edge] is like plunging into a river in spate' (*London Orbital* 378), '[t]he fording of the M1 is a big moment' (*Edge of the Orison* 20). In the above passage, the 'blam-blam-blam' of passing cars shares aural space with the similarly repetitive tapping of the woodpecker. Lastly, these observations take place at twilight, recalling an elegiac tradition of gloaming reflections mourning passing. Double visioning is thus, in this instance, both a reflection on the sensory experience of walking this stretch of road – incorporating sounds alongside sights – and a recognizably literary device. As Sinclair makes clear in the following paragraph, this perspective is predicated on walking: 'The walker "goes back", forgets himself (or herself). A pre-visionary condition, in which it is possible to let go of the present and access an older narrative, a secret garden or enchanted wood' (*London Orbital* 138).

The road itself is also periodically treated in a bucolic register. The interchange of M23 and M25 'blend[s] the picturesque with the functional': 'A three barred safety fence replaces the five-barred gate as somewhere to lean, chew a stalk of grass, watch the road. A heat-singed motorway palette encourages contemplation; dark greens and burnt browns disappearing into a range of recessive silvers and blues' (375). Reflections whilst leaning on a convenient gate (nodding to the poetry of Gray, Hardy, and many others) aestheticizes the motorway itself. Its description in pastoral terms debases their efficacy and makes them comic. The road itself, which is overlooked or even positively elided in the promotional image of London's edgelands, is reinscribed in this reckoning. The process of aestheticization itself is thus interrogated.

Sinclair also formulates narratives of connection between the past and the present by remembering neglected histories. His texts possess what Alistair Bonnett calls a 'restitutive impulse' by which they aim to forge 'bonds of solidarity and empathy with earlier generations of radicals' (57). It is by walking that Sinclair forges bonds with a dissenting history, and thus reinstates it in the landscape traversed. As he approaches the interchange between M25 and M4, Sinclair's thoughts turn to the counter-cultural or avant-garde poetry scene of the 1960s and 70s, of which he was part. He then goes on to link that 'decade of samizdat publication (1965–75)' to 'the outpourings of the Dissenters (Levellers, Diggers, Ranters) in the years after the English Civil War (1646–56)' (225–6). Sinclair inserts his contemporaries' poetic voices into a dissenting history, but also into the landscape itself: 'I began to superimpose those radical songs … on to the Colne Valley, our march towards Heathrow.' An association of place and poetry is the result: 'The sounds of the road, as the M25 approached the tangled interchange with the M4, were compulsive; as complex and as many-voiced as a … poem. A sound that was its own score' (227). Aural perception of environment and personal or cultural resonance coalesce here. This passage, in isolation, may appear to be a particularly subjective imposition of a particular narrative onto an arbitrary location (Sinclair's verb is 'superimpose'), and indicate a tendency directed towards reverie more than alertness (the sounds of the road are 'compulsive'). However, contextual passages make clear that these connections are not only imposed on, but also emanate from the (forgotten) histories of the particular locations passed through.

For Sinclair, 'English dissent is a residual presence [in the landscape]: Cromwell and his associates, soldiers, bible-punchers, republicans' (*Edge of the Orison* 163). Attention is again extensively conferred on seventeenth-century dissenting traditions when Sinclair and his co-walkers reach St George's Hill near Weybridge. 'If psychogeography is the theme', he writes, 'Weybridge has it – well disguised, screened by foliage, always present' (*London Orbital* 299). The project, here as elsewhere, is to pull aside the screen to reveal the histories of place that are neglected but nonetheless 'always present'. It is to rediscover 'residual presences'.

The walk 'over St George's Hill, and on towards Cobham Heath' is a retracing, 'following in the steps of Gerrard Winstanley and the community of Diggers, in the period after the English Civil War' (300). The language deployed to describe the walk here recalls the link between ritual and exorcism assessed above. St George's Hill, Sinclair notes, is a 'place of pilgrimage' where Winstanley and the Diggers, calling themselves the True Levellers, asserted that the 'land belonged to those who worked it'. This location is a 'sacred place for dissenters. Common land was developed as a private estate in 1911. The hurt remains' (305–6). The ritual purpose of Sinclair's 'pilgrimage' is to reassert the 'sacred' character of a place; to exorcise subsequent privatization, and reinstate a counter-cultural inheritance. If Sinclair's religious references – here to Cromwell and dissent, but elsewhere also to rituals, ceremonies, and histories associated primarily with Catholicism – seem contradictory, the quality that they share is recusancy. In varying forms, these very different lineages signify for Sinclair a refusal to submit to the authority of the established Church of England. They provide a composite model for his own similar (though secular) antiauthoritarian project.

In these passages, Sinclair aims to affirm connections between his own walking practices and that period when '[s]plinter groups, fanatics and visionaries of every stamp, took to the roads' (306). Links are established between the experiences of seventeenth-century radical walkers and Sinclair's own traverse of St George's Hill. The text records how those earlier pedestrians 'were thought to be conveyors of intelligence, spies, plotters, heretics', and describes Restoration-era harrying of travellers: 'Dissenters were persecuted. Justices of the peace harassed and imprisoned vagrants. England was brought to that happy state where those who roamed – without good reason, without passports or permissions – were liable to be defined as being out of their wits, Tom O'Bedlams' (308). The alignment of these walkers 'out of their wits' and Sinclair's own 'mad walking', well established by this point in the text, is clear. It is precisely these dissident associations that *London Orbital* construes and plays off in its casting of walking as an activity possessing radical potential. The associations are reaffirmed in the surveillance and harrying that Sinclair is exposed to in his attempt to pass through the private estate currently occupying St George's Hill, the approach to which 'is orthodox Surrey: a public road that, quite suddenly, isn't'. He records the relentless supervision: 'You are allowed to walk half a mile between security shakedowns. Slow moving Fiestas are on constant patrol. CCTV cameras, panning restlessly, alert the monitor jockeys' (302, 308). This security infrastructure stands in metonymically for attempts to secure the present status of this place as a *private* estate that Sinclair's description of its history as a piece of *public* 'common ground' serves to undermine. The remembering of an elided history of place, reinscribed in the process of walking, loosens the bounds of its current condition. An alternative account of place is thus offered alongside the official rhetoric of the zones that Sinclair's text is concurrently working to undermine. These two facets of Sinclair's ambulatory engagement with the edgeland landscape – aesthetic and historical – complement and co-depend upon one another.

## Walking and Writing

Walking, then, is thematically significant for Sinclair. Indeed, it provides crucial underpinning for both the aesthetic and historical interventions of *London Orbital*. However, cultural geographer Tim Edensor has cited Sinclair's texts as an example of the way in which walking narratives 'curiously decentre corporeal, sensual interaction with the material world'. In 'foregrounding the metaphor of walking as narrative inscription', he suggests, 'the affective, sensual dimensions of walking are apt to disappear' (136). I argue here that this is *not* the case. Hayden Lorimer and John Wylie contribute the following to the debate on walking narratives:

> Whenever and wherever you walk, [the] commanding thesis of integration – of an existential continuum of mind, body, and world – dis-integrates with almost every step, every dis-location that is, by which you seem to be less the flowing sentences of the landscape and more its jarring punctuation, all commas and colons and question marks. (10)

This might seem to concur with Edensor's assertion. Indeed, one of the things that Lorimer and Wylie appear to be saying is that the very mobility of walking undermines the coherence of narrative readings of landscape and place: dis-location leads to the dis-integration of connections between self and world that is represented in flowing sentences. However, the jarring punctuation – the commas, colons, and question marks of stopping and starting, of changing pace and of unfolding landscapes as one walks – are still a literary, or even more basically, a textual feature. What might be taken from this remark is not a disregard for narrative and stylistic analysis, but a more considered dovetailing of geographical and literary concerns.

If, for Sinclair, the rhythm and pace of walking facilitate a certain kind of engagement with place and therefore produce a particular kind of text, the relationship is reciprocal: the text aims, structurally and syntactically, to reproduce the experience of, and perceptions emanating from, the practice of walking. In a return to the 'syntax of London' with which this chapter opened (*London Orbital* 158), attention is now turned to more literal modes of punctuating in order to argue that Sinclair's style attempts to recuperate – in the text's texture, its formal properties – the processes of walking on which it reports.

Digression opens an interface between the structural and syntactical aspects of Sinclair's walking and writing in *London Orbital*. After over 100 pages of preamble, exposition, and the walk up the Lea Valley to the starting point for the M25 circuit at Waltham Abbey, Sinclair reports that 'the walk proper' begins here: 'No detours. No digressions' (125). However, predictably enough and only two pages later, Sinclair and Renchi do digress from following the motorway when they discover a path marked with granite pillars laid out through Cornmill Meadows behind the Abbey:

> The walk we are offered, strategic planting designed to flatter a diminishing perspective, is very seductive: the distant prospect, at the end of a closely mown

> avenue, of another pillar. It carries us away from the M25, but it's irresistible. A grassed extension of the Lea. A spirit-path that runs in parallel with the Navigation, the Cornmill Stream. (127)

The digression from the planned route for the day's walking is emphasized rather than obfuscated in Sinclair's description of the incident, and acts as something of a thematic marker for the text itself. Sinclair's accounts of these places, his undercutting of conventional narratives concerning them, and his relentless proposal of counter-narratives, act metaphorically as forms of digression. Similarly, in a motif developed throughout the text, the planned route is digressed from when contingent experiences intervene. If digression here signifies various forms of subversion, it is also described in a particularly idiosyncratic, digressive style that critics have often emphasized in their responses to Sinclair's writing.

Narratological and stylistic accounts of Sinclair's texts gravitate towards the governing principle of parataxis that they display. Brooker has described the 'obsessive, digressive movement that drives his sentences and waylays his narrative' (*Modernity and Metropolis* 100); whilst Baker has found that 'the brilliance of language at the level of the sentence or line ... creates a "semantic drag", at the expense of narrative propulsion in Sinclair's prose fictions' (*Iain Sinclair* 124). Barry describes the prose as 'disjunctive', consisting 'of brief, sawn-off sentences which are placed end-to-end without connectives. The structure is relentlessly paratactic' ('You can get it from the street' 45–6). Sinclair's writing 'consists of an accumulation of disjunctive, serial "hits" on various targets (or sometimes repeated hits on the same target), rather than an unfolding line of argument. Each hit is peculiarly rounded-off and self-contained, and hence often quite disruptive of the larger discursive unit which contains it' (46). Barry goes on to develop a narratological model wherein teleological and entropic drives compete:

> The T[eleological]-Drive, or forwarding momentum, is counteracted by the E[ntropic]-Drive, which seeks to stop the flow and arrest the attention on a single moment. The E-Drive is located in the local force and power of individual passages (moments, phrases, scenes, and so on) – 'purple' passages as they used to be called, which can be so vivid as to eclipse the larger direction and purposes of the work, making it difficult for the work as a whole to resume its onward movement. (50)[10]

Applied to the passage quoted above, this critical model describes the way in which subsequent clauses and sentences reconceive the scene, adding layers of description without moving forward: the path's 'diminishing perspective' is recapitulated in relation to the 'distant prospect' of the pillar in the distance, then becomes a 'grassed extension' of the River Lea, and finally a 'spirit-path'. In this digression, modulation tends away from actual and towards metaphorical

[10] In more traditional narratological terminology, this effect could be described as 'descriptive pause (maximum textual space, zero story time)', or 'slow-down or stretch (textual space greater than story time)' (Bridgeman 58).

landscape, as well as away from a planned and guided route towards an alternate reading of place. Structural and syntactical techniques – the privileging of entropic over teleological drives – are deployed (like walking itself) to slow down progress and refocus the environment passed through in the process. If, for Sinclair, the promotional material for London's edgelands is 'de-fanged, homogenised', then his own paratactic style aims to reinvigorate the representation of these zones (*London Orbital* 262).

Narratological theorists often employ a suggestively spatialized language in their descriptions of plot and narrative, that chimes with the point being made here: Roland Barthes's 'hermeneutic code' employs 'equivocation', 'snares', and 'false replies' to 'set up *delays* (obstacles, stoppages, deviations) in the flow of the discourse' and maintain a tension between propulsion and stasis (75); Peter Brooks finds that 'plot itself stands as a kind of divergence or deviance', and that '*détour*, the intentional deviance, … is the plot of narrative' (291–2). These figures are pertinent, as is seen here, in a discussion of Sinclair's entropic narrative style at both structural and syntactical levels. A number of critics have theorized Sinclair's style in relation to the representation of space and place: Baker, for example, suggests that Sinclair 'propose[s] a "poetics of space" in terms of [his] site-specific texts', but also in a textual form that 'itself becomes spatial rather than linear' (*Iain Sinclair* 42). However, less attention has been devoted towards the *perspective* from which Sinclair's spatial texts are written, and which they mark stylistically. Neither is there a critical consensus on this matter when it is approached incidentally: Brooker finds that 'Sinclair's wayward rocketing prose, bursting with meaning, is launched *from a standing position*' (*Modernity and Metropolis* 101); whilst Joe Moshenka finds that, in Sinclair's paratactic delivery, '[t]he ideas flash out in quick and anarchic succession, much as the perceptions crowd in to the mind *of the walker*' (n.p., my italics). The difference between these critical opinions is instructive – Sinclair's style might seem to hover between a static, standing and a mobile, walking perspective. In fact, his writing presupposes and recuperates a pedestrian perspective in the space of the text.

The pace of walking constructs a suture between situated and mobile perspectives. Sinclair is certainly attentive to this issue. In *Edge of the Orison*, he writes that once he has gotten 'into my stride', that is, achieved a certain walking pace, 'a comfortable momentum is achieved, at which certain details are registered, but the freewheeling mind doesn't drag like an anchor; you are not caught up in the anticipations of arrival' (106). Progression and dalliance, by virtue of the pace of walking, are held in tension. In critical discourses, the representation of place (both visual and textual) is often associated with situated, if not always static, perspective; whereas the experience and practice of place is commonly associated with processes of moving through and consequently with a mobile perspective. Walking, I contend, collapses this critical dichotomy.

This occurs primarily by virtue of its comparably slow pace of movement. Solnit writes of a walk as 'sewing [space] together into a continuous experience', and suggests that for the walker, 'landscape moves by as a gently modulated continuity' (4, 174); Jarvis finds that '[t]he pedestrian's experience of the world is of a slowly but continuously changing field of appearances' (68); and Ingold suggests that when we walk 'things fall into and out of sight, as new vistas open up and others are closed off', and that as a result 'the structure of our environment is progressively disclosed' (*Lines* 87). Small, incremental movement and shifting perspective approximate the still, the framed, without abandoning continuous and ongoing change and revision ('a gently modulated continuity'). As a result, a literature of walking, Gilbert argues, is differentiated from static models of landscape description in that 'it takes as its subject not a landscape, an essentially static entity, but an experience' (8). The perspective that informs Sinclair's walking texts is one that neatly expresses the interdependence of representation and practice in the production of literary landscapes.

Theorizations of a walking perspective as representing not a landscape itself but an experience of that landscape intersect with several critical responses that emphasize the experiential register of Sinclair's texts and their stylistic characteristics. For James, 'Sinclair's cryogenic sentences unfold to retain something of the unfolding volatility of spatial experience' ('As He Stalks' 157); and for Perril, 'Sinclair's generation of velocity through stunted sentences, often composed of one word … presents events as a myriad of discrete flashes as yet unsorted by any presiding intelligence' (332). The terms of both of these accounts resonate with walking. 'Unfolding' links the paratactic layering of Sinclair's sentences to the 'gently modulated continuity' of pedestrian perception. A 'stunted style' also goes some way towards reproducing the process of 'sorting' a 'myriad of discrete flashes' that the walker's slow movement predicates. Wolfreys finds that Sinclair's texts are 'radically *graphic*' because they operate 'via synchronic revelation' (that is, they comprise paratactic layers of perception). His mode of writing, 'in being irreducible to a single order of signs architectonically arranged, does not merely describe or narrate an event, it effectively stages that event through the multiplicity of signs configured as a simultaneous transmission' ('Londonography' n.p.). The 'event' that these passages narrate, then, is surely walking itself.

Another passage from the same Waltham Abbey section of *London Orbital* serves to illustrate this mode of representing process in action, of reproducing walking in the poetics of the text:

> A day of locked gates. The abbey was off-limits, a funeral. So we wandered through the orchard, the monastic reservation, circumnavigating drained fish ponds. … A residue of retreat, monasticism, is still present in whatever remains of the original layout; measured avenues, monuments to the godly or powerful, warm red bricks. But, play the empathy game as much as you will, you can't escape the song of the road, the mantra of transit. A perpetual cycle of auto-prayer. (126)

Again, the passage emphasizes digression and departure from the planned route or agenda – as the gates are locked (the abbey being in use for a funeral), Sinclair and Renchi 'wander', that is, walk aimlessly; the 'circumnavigation' of the fish pond stands in metonymically for the circuit of the M25 that they ought to be beginning at this point. The two lists present unfolding perceptions of a walking perspective, detailing firstly the features passed – 'the orchard, the monastic reservation, the drained fish ponds' – and secondly registering a shifting perspective ranging from the far to the near – 'measured avenues, monuments to the godly or powerful, warm red bricks'. The experience being recorded is not a solely visual one: this walking is accompanied by the ever present sound of traffic on the motorway. Clauses that describe and clauses that interpret are interspersed to reproduce the interplay between perceiving and conceiving that the pace of walking entails and encourages.

The paratactic style and favouring of the halting entropic drive also serve to facilitate the double visioning aesthetic detailed above. The 'residue' and 'remains' of older incarnations of place are patterned onto its current guise: in this instance, the monastic chants that might have been heard here in another era have become 'the song of the road, the mantra of transit. A perpetual cycle of auto-prayer'. This sequence of layered metaphors emphasizes discrepancy alongside likeness. Brooks has written that whilst metaphors have usually been understood to rely on resemblances, 'modern treatments of the subject have affirmed equally the importance of difference included within the operation of resemblance, the chief value of the metaphor residing in its "tension"' (280). In this example, it is the disparity as much as the similarity between the two components of the metaphor that informs Sinclair's meaning: rather than expressing consonance, old and new characteristics of place disrupt one another here. It is the slow pace of walking, recuperated and reproduced in the stylistic texture of the passage, which allows for the aesthetic linking and disjuncture established between past and present. It is by these means that Sinclair's text steers a course between the imperatives of recording experience and formulating a narrative representation.

Shklovsky writes that techniques of defamiliarization aim 'to make forms difficult, to increase the difficulty and length of perception because the process of perception is an aesthetic end in itself and must be prolonged' (12). In comparison to vehicular modes of transit, the act of walking has this prolonging effect, extending the 'length of perception' with which a landscape is viewed and experienced – the kinship between the walking and defamiliarizing is emphasized further in Shklovsky's observation that the latter often takes the form of 'disordering rhythm' (24). If the motorway landscapes of the text are defamiliarized, it is because they are experienced in an unfamiliar way: on foot. Rather than distancing the text from the experience that produces it, defamiliarization therefore actually works to offer a closer approximation of that experience. If the road, as Sinclair describes it at the text's outset, 'acts as a prophylactic between driver and landscape' (*London Orbital* 3), then walking reinstates a fuller engagement in slowing the process down from passing *over* to passing *through*. Similarly, and coterminously, the digressive, entropic style of Sinclair's writing – at both syntactical and structural levels – is always twisting away from its end, taking the long route. It is by these means that memory is reintroduced into a forgotten place, and that the defamiliarizing effect of the refusal of coherence between past and present finds it force.

# Chapter 4
# Engaging: Robert Macfarlane

'There is a crisis of representation and exposition in landscape aesthetics', Ken Worpole declares in *The New English Landscape* (2013). 'Lofty statements on the true and the beautiful are increasingly resisted as the particularities and historical complexities of local and regional topographies – especially those of former industrial areas – are evaluated anew' (10). The aesthetic shift that Worpole describes – from the pristine to the blended – is a response to material changes in that landscape itself: 'The hard and fast distinction between town and country is no longer tenable in many parts of Britain today, especially as road building, housing development and retail parks have extended deep into former rural terrain, whilst at the same time agriculture itself has industrialised.' He declares that 'the requirement to interpret and re-evaluate contemporary landscapes – especially those which resist traditional categories of taste – is therefore vital' (12). Worpole's writing is an intervention in its own right (its titular focus on 'newness' emphasizes the programmatic component) undertaken via a mapping of changes already underway. Thus, he notes that artists' colonies have previously been 'located as far from industry, commerce and metropolitan life as was possible', but 'today the romance of the remote is no longer a part of landscape aesthetic. If anything, the opposite is the case'. Indeed, we are currently seeing 'a growing emphasis on the vernacular and familiar as equally worthy of artistic interest' (13).

If there has been a 'crisis of representation' in landscape aesthetics, it is one that has been attended by a significant body of writing in the last 10 years. Indeed, in 2008 Patrick Curry asserted that 'a renaissance in British nature writing is now underway' (26). Worpole is not the only writer to deploy diagnoses of crisis as the lever for comparable declarations of novelty of approach. The 'new nature writing' of this period has been at the forefront of attending to contemporary landscapes – refusing to exclude from 'nature' those places that bear the marks of human intervention – and of developing new perspectives from which to account for them. The appellation was coined in a 2008 issue of *Granta* magazine on the subject of 'The New Nature Writing', in which Jason Cowley's editorial letter characterizes contributions as 'new ways of seeing' nature from 'writers who approached their subject in heterodox and experimental ways' (10–11). This description in fact trails a long legacy, encompassing both accounts of attention and perspective emerging out of the phenomenological work of Maurice Merleau-Ponty and others, and the assertion that 'landscape is … a way of seeing' of cultural geographers such as Dennis Cosgrove (55). New nature writing calibrates the two (hence its value to this study).

For many commentators, Robert Macfarlane has been the pivotal figure around whom this movement has recently turned. If foregoing chapters of this book have addressed writers who, to varying degrees, represent the social – the human and settled – aspects of place, then the environmental, non-human facets implicit there now become the focus of attention. Moreover, one of the defining features of Macfarlane's new formulations of nature writing approached here, and of nature writing as a genre, is the exploration of humans' relationship with the non-human environments they inhabit. This often takes place through the employment of hybrid or generically ambiguous literary forms that encompass (and interleave) both experiential and cultural accounts of place.[1]

Though scholarly response to the rejuvenation of non-fiction nature writing and to the connected growth of a body of work addressing itself to ideas of wildness is in its infancy, those treatments that have appeared are instructive in identifying key figures and defining important features of the emerging movement. Stephen E. Hunt focuses on Mark Cocker, Roger Deakin, Richard Mabey, and Robert Macfarlane as a central group of new nature writers who all published important texts in 2007, and whose 'respect and mutual encouragement of each other's work' suggests something of a cooperative coterie (70). The 2007 texts Hunt refers to are Cocker's *Crow Country: A Meditation of Birds, Landscape and Nature*, Deakin's *Wildwood: A Journey Through Trees*, Mabey's *Beechcombings: The Narratives of Trees*, and Macfarlane's *The Wild Places*. All four also contributed pieces to the *Granta* 'New Nature Writing' issue. If Hunt's article testifies to literary scholars' interest here, attention on the part of cultural geographers is demonstrated by David Matless's review of the same four texts, finding that '[t]heir narratives perform through interactions of human subject and nature subject matter, with reflexivity concerning past conventions of nature writing and attention to uncertain boundaries of culture and nature' ('Nature Voices' 178–9). Like Hunt, Matless notes 'a web of connection and citation between these nature voices, lending logic to their grouping' (179).

The debates that these critical pieces stage, when taken together, is one about the varying modes of 'wonder' that these new nature writing texts embody. Many of the past conventions alluded to by Matless derive from the long tradition of natural history writing that these writers undoubtedly engage with reflexively. Matless tends to critique new nature writing – Macfarlane in particular – for the degree to which 'wonder' as curiosity ('quizzical observation and scrutiny')

[1] Other definitions of nature writing attempt to specify genre. For example, Timothy Clark: 'Nature writing continues to be used as a term to describe a kind of creative non-fiction associated with usually meditative accounts of natural landscapes and wildlife' (5). Raymond Williams's *Keywords* includes an essay on 'Nature', which has amongst its definitions 'the material world itself, taken as including or not including human beings', but from the late eighteenth century has more specifically signalled 'the "countryside", the "unspoiled places", plants and creatures other than man. The use is especially current in contrasts between town and country: nature is what man has not made, though if he made it long enough ago – a hedgerow or a desert – it will usually be included as natural' (219, 223).

about nature, the hallmark of natural history, has been replaced by 'wonder' as marvelling at nature (185). Hunt responds to the same feature – the feature that perhaps most readily characterizes new nature writing – more positively. For Hunt, acute awareness of their positions as formulators of narratives of nature 'liberates all four authors to engage with their themes in a way that avoids spurious detached objectivity while retaining a healthy respect for the otherness of their subjects' (71). Anna Stenning and Terry Gifford have accounted for this personalized approach, 'a familiar pattern of the sensitive and informed individual's encounter with nature', by drawing attention to its inheritances from twentieth-century American nature writing. This register contrasts a British tradition 'less spiritual and more concerned with natural history, perhaps less internally and more externally focused' (1). In interview, Macfarlane has also drawn attention to this tonal influence: 'between 1950 and 1990 North American nature writing cycled through the phases, and underwent the same mutations, that we are now seeing repeated almost precisely ... in "British" Anglophone writing of a similar kind' (qtd. in Stenning 79). If Macfarlane has been the writer most often critiqued for the apparent shortcomings of this new way of writing nature, he has also been its most self-aware practitioner and articulate spokesman. For this reason, his texts offer a test case for an assessment of this broadly adopted new aesthetic.

In a foreword to *A Wilder Vein*, a collection of new literature of place and nature published in 2009, Macfarlane indicates his sense of the new writing's defining characteristics: 'over the past decade or so in Britain and Ireland, a body of written work has emerged that has tried to generate a polytheistic language for the wild: from wildness as a state of land to wildness as a state of mind' (vii). If the *Granta* 'New Nature Writing' issue accommodated a diverse range of authors working in a variety of modes, *A Wilder Vein* encompasses an even wider sample of new nature writing with contributions from new writers and established figures who have turned towards this subject matter. Macfarlane has found himself in the position of being increasingly called upon to critically define the movement in which he is a leading figure. As a result, he has been influential in attempts to formulate something like a canon of nature writing. 'Common Ground', his series of essays in *The Guardian* in 2005, argued for the reissue of 'classic works of nature writing from England, Scotland, Wales and Ireland', that would signify 'a series of local writings, which concentrated on particular places, and which worked always to individuate, never to generalise' (n.p.). This formation shares Worpole's particularizing drive and attentiveness to local topographies, with which this chapter opened. Macfarlane's subsequent forewords to numerous reissues have contributed further to this process of reclamation, forming a loose group of precursors into a nascent canon.[2] His work signifies most clearly both the

[2] The reissues for which Macfarlane has written introductions include: J.A. Baker, *The Peregrine*; John Christopher, *The Death of Grass*; John Stewart Collis, *The Worm Forgives the Plough*; Ian Frazier, *Great Plains*; Jacquetta Hawkes, *A Land*; W.H. Hudson, *Adventures Among Birds*; Richard Jefferies, *Nature Near London*; John Muir, *My First Summer in the Sierra*; Tim Robinson, *Stones of Aran: Pilgrimage*; Nan Shepherd, *The Living Mountain*; and Edward Thomas, *The South Country*.

inheritance contemporary writers have drawn from a rich history of nature writing and the new departures that indicate the 'genuinely contemporary movement' that he himself hails it to be (vii).

Macfarlane's identification of a shift from wild states of land to wild states of mind suggests a tension between externalist (documentary) and internalist (testimonial) approaches to engaging with wildness, nature, and place. Macfarlane's own position on a continuum between these two modes is perhaps indicated in his recent assertion that in poetic accounts, 'the "I" gets aggressively dismantled'; whilst '[t]he lyric "I", the first-person witness and narrator', is 'central to [contemporary] prose writing about landscape' ('Meshed and weathered' 14). He surely has his own prose writing in mind here. Readings of these texts demonstrate how they bear out the convictions that his editorial and curatorial work voices. The commitment to 'wildness as a state of mind' is evident in three evocative books: *Mountains of the Mind: A History of a Fascination* (2003), *The Wild Places* (2007), and *The Old Ways: A Journey on Foot* (2012). They form, in Macfarlane's own terms, 'a loose trilogy about landscape and the human heart'. This commitment informs the testimonial approach to engaging with nature and place taken up in all three, though the mutuality of the relationship between non-human environment and human heart is indicated in the author's note to *The Old Ways*, where Macfarlane alludes to 'the subtle ways in which we are shaped by the landscapes through which we move' (xi). This reciprocation informs all of his writing. The three books of the trilogy are treated in this chapter, with *The Wild Places* as the pivotal text dwelt on longest.

## Language, Description, and Narrative Engagements

Preliminary to an extended treatment of these texts, it is necessary to make an evaluation of the investments made in processes of engagement with nature *per se*. Like preceding sections of this book, this chapter is titled with a gerund naming a mode of interaction between writer and place. The potential problem of what constitutes 'engaging' with place is one that is addressed by several contributors (including Macfarlane) to another recent collection of essays, *Towards Re-Enchantment: Place and Its Meanings* (2010). The editors of this volume, Gareth Evans and Di Robson, ask contributors to write on the personal resonance of a specific location. In their introductory remarks, they allude to the agenda that informs the volume: 'At a time when the multiple alienations of modern society threaten our sense of belonging, the importance of "place" to creative possibility in life and art cannot be underestimated' (n.p.). This traditionalist sense of the threat posed by encroaching modernity carries Weber's formulation in which 'rationalization' and 'intellectualization' have led to the 'disenchantment of the world' (155). As such, *Towards Re-Enchantment* can be situated in relation to a broad contemporary trend: Joshua Landy and Michael Saler suggest that there are currently 'a variety of secular and conscious strategies for re-enchantment' by which 'modern intellectuals and creators have put together a panoply of responses to the Weberian condition, offering fully secularized subjects an affirmation of existence that does not come at the cost of naïveté, irrationalism, or hypocrisy' (2). Others are more circumspect: Neal Alexander and David Cooper observe that 'enchantment may actually impede rather than enable reflexive engagements with place and mystify the cultural processes whereby landscapes are constructed and understood' (4).[3] In the concept of re-enchantment and the potential for (re)connection and 'belonging' that it offers, a residual Romantic inheritance might be felt. Enchantment itself is a specific form of engagement particularly attentive to the emotional qualities of place and nature. It therefore posits an alignment or continuity between self and place that is not entirely unproblematic. Some contributors to the volume simply take 'enchantment' and 're-enchantment' to be synonyms, but the latter term does carry an extra resonance of restoring to place that which has been lost. In these ways, the premise on which the collection is predicated introduces questions of projecting narratives and meanings onto or into places (the volume's subtitle presents 'place and its meanings' as curiously separable entities divided by a conjunction), and several contributors recognize the dangers attendant therein.

In his contribution, Iain Sinclair, for example, notes the potential difference between assigning and accepting the meaning of place. In a passage already addressed in this book (see Chapter 3), he writes that '[m]aking yourself ready to accept the dictation of place is the first requirement: and then the unexpected, that wished-for second consciousness, will happen' ('Water Walks' 19). The 'second consciousness' described is ambiguous, but it is clear that it emerges from an effacement of one's projections in a readiness to accept place's dictations – the role of the writer here is reduced, seemingly, to that of scribe. It is far from clear that Sinclair's texts always conform to this model, but it certainly is a concern that informs the new nature writing approached here. Richard Mabey's contribution to the volume addresses East Anglia and focuses on the region's relationship to water and the sea. He finds that '[i]f there is a defining East Anglian narrative, a story that shapes the region's sense of its own identity, it is the parable of inundation' ('On the Virtues of Dis-Enchantment' 30). A similar ambiguity to that of Sinclair's construction can be felt here. The region's character clearly emanates from its physical geography (the specific topography of the area results in regular inundation or flooding), but to describe this as a 'parable' suggests an additional allegorical significance in accounts of inhabitants' relationship to place that is more actively inscribed. In any case, it is clear that narrative and narration have a role to play in engaging with nature and place. This informs new nature writing in a variety of ways.

Robert Macfarlane's piece in *Towards Re-Enchantment*, 'A Counter-Desecration Phrasebook', is also attentive to the issue of ascribing meaning and deals centrally with the question of the relationship between language, description, and place.

---

3 The 'disenchantment thesis' is also interrogated in Jane Bennett's *The Enchantment of Modern Life* (Princeton: Princeton University Press, 2001).

Language and, more specifically, naming are central to much of his work. This piece is concerned with a visit to the Isle of Lewis in the Outer Hebrides and what is revealed by the extent of the 'Hebridean Gaelic lexis of peat and peat-moorland' used by local inhabitants there. In a technique that also informs *The Wild Places*, a shift in perspective over the course of the text is deployed here. Initially, the problem of assigning meaning to place is confronted from a distanced, outsider's perspective. The account opens with the view of Lewis's moorland from the plane that brings Macfarlane to the island: 'Seen for the first time, and especially when seen from altitude, the Moor can easily resemble a *terra nullius*, a nowhere-place … at first glance the Lewis Moor appears distinguished only by its self-similarity' ('Counter-Desecration Phrasebook' 108). As a result, interpretation is problematic: 'This is a region whose apparently undifferentiated expanse seems either to return the eye's enquiries unanswered, or to swallow all attempts at interpretation. [This landscape] confronts us with difficulties of purchase (how to anchor perception in a context of vastness) and evaluation (how to structure significance in a context of uniformity)' (108). Apparent featurelessness inhibits engagement.

In sharp contrast, the insider's perspective is marked by an extreme linguistic specificity that demonstrates an intimacy with place that is valued highly in Macfarlane's own work. The text describes how Finlay McLeod, a Hebridean friend of Macfarlane's, has compiled a peat glossary of the extensive range of terms by which those who live and work here discriminate the many forms in which this substance is found when the landscape is perceived and felt up close. The same respect for the 'exquisite exactitude' of Hebridean Gaelic registers in *The Old Ways* (164). For Macfarlane, 'a lexis of such scope and such exactitude' testifies to an 'intense relationship of labour' between inhabitants and land; but furthermore, '[t]hat this language should also so readily admit the poetic and the metaphorical to its designations is testimony to the aesthetic relationship between the inhabitants of the Western Isles and their land' ('Counter-Desecration Phrasebook' 109). The assignment of meaning and the perspective from which it emanates are dependent on the nature of the engagement with place and nature. The recognition of these factors informs Macfarlane's texts in terms of both the modes of engaging detailed there, and the narrative positions from which that engagement is reported and described. Indeed, Macfarlane's writing is always shifting from reportage towards description. Like the islanders that he is discussing in this piece, his own descriptive aesthetic is one that relies on linguistic specificity. The mannered introduction of particular pieces of vocabulary characterizes his work. 'Wildness' is perhaps the most important word in Macfarlane's work: as subsequent analysis will show, he is constantly at pains to distinguish it from 'wilderness' and to home in on its etymological significance. This revival of certain terms – and the tendency to dwell on their etymologies – suggests something of the recuperative drive that situates his work within a project of *re*-enchantment.

Indeed, re-enchantment in language runs counter to desecration, as the title of the piece makes clear. This is where the political efficacy of Macfarlane's work lies. As he remarks elsewhere, 'naming … operates often as a mnemonic or preservative' (Introduction to *Nature Near London* xx). As Jos Smith observes, alertness to terminology 'seems to offer more than just nomenclature. It offers an insight into careful acts of attention, and alludes to personal and community narrative, perhaps even the beginnings of – with a small "c" – culture as brokered between land and people, land and community' (6). Linguistic richness signifies a mark of value and thus a first step towards preservation. This informs a number of features of Macfarlane's texts. *Holloway*, a recent collaborative piece treating the now-disused sunken lanes scored into southern English landscapes by centuries of use (i.e. hollow-ways), opens by revelling in the historical variety of the term, laid out on the page with the line breaks of poetry:

> Hol weg.
> Holwy.
> Holway.
> Holeway.
> Holewaye.
> Hollowy.
> Holloway.
> (2)

Similarly, in *The Old Ways*, Macfarlane revels in a list of historical and regional terms for 'path' – 'trods, leys, dykes, drongs, sarns, snickets' (and many more) – as 'they become a poem or rite' (13). If this kind of inclusion bespeaks an aesthetic engagement and forgoes explicit explanation, Macfarlane elsewhere offers detailed etymologies and teases out their implications. Significantly, *The Old Ways* is also provided with a glossary (all three texts of the trilogy afford a bibliography and other textual apparatus, with similar ends in mind). Macfarlane suggests that textual records seek to 'archive in language what [is] vanishing, or to recreate in language what [has] already gone' (*Old Ways* 221).

Crucially for Macfarlane, the language of Lewis's inhabitants demonstrates the coalescence of the ecological and the poetic. Intimate engagement with a landscape results in 'a lexis so supplely suited to the place being described that it fits like a skin. Precision and poetry coexist here: the denotative and figurative are paired as accomplices rather than antagonists' ('Counter-Desecration Phrasebook' 110). The same problem that Sinclair and other contributors to *Towards Re-Enchantment* recognize is negotiated here by Macfarlane in a fashion that allows for the projection of meaning without the appropriation of place's agency on the part of the writer. This ability emerges from an intimacy with place and a kind of linguistic environmental determinism. Following from his discussion of Lewis, Macfarlane asserts that 'language does not just record experience stenographically, it produces it. Language's structures and colours are inseparable from the feelings we create in relation to situations, to others, and to places.' In short, '[c]ertain kinds of language can restore a measure of wonder to our relations with nature' (118). This argument is comparable to that made in another arena by Jonathan Bate, who argues that 'the poem might re-enchant the world. It can only do so if it is understood as an experiencing of the world, not a description of it' (*Song of the Earth* 167). The difference between these two statements lies in Macfarlane's stronger sense of 'description' and its role in experience: in his thinking, the two coalesce. Macfarlane's model for engagement recognizes the literal registers of 'en*chant*ment' where singing or chanting, representing in language, plays a key role. In this thinking, an attempt is made to detach wonder from its association with the miraculous. Here, wonder does not signify a fanciful departure from more circumscribed models of participation, but in fact emerges out of them. This transition is crucial to Macfarlane's own literary style. 'A Counter-Desecration Phrasebook', written after *The Wild Places* received criticism from some quarters (for example, Matless, above), seems at least in part to offer a response. This fact perhaps informs the theoretical tone that replaces the more effusive style of Macfarlane's earlier nature writing. It presents wonderment as a viable and useful model for engagement.

Language and description are important to Macfarlane's engagements with nature, but it is perhaps narrative that is the single most significant feature there. In revising narratives of nature and place, Macfarlane's work is, in part, made possible by important forerunners in this area. Ken Worpole's contribution to *Towards Re-Enchantment* chimes with his work in *The New English Landscape.* His piece on Essex, 'East of Eden', is the most explicit of the contributions in formulating a new set of values for landscape, and in providing a rationale for doing so. For Worpole (and, as will be seen, for Macfarlane), this county's coastline is a paradigmatic model for the newly ambiguous landscapes that operate outside habitual aesthetic categories: 'This Essex coastline embodies a melange of the maritime and the industrial, the defensive and the arcadian, much of it now redundant.' These places' 'potent mixture … confounds and dismays contemporary landscape aesthetics' (65). It is the recalcitrance with respect to established ways of thinking and the liminality of these spaces that demands a revised aesthetics – Essex is chosen, in part, as it is literally on the eroding, eastern edge of the country. The response of recent nature writing has been to consider the ways in which literary narrative forms might respond to the need for a revised aesthetics and poetics.

Richard Mabey's writing is crucial to opening up new narratives of nature that Macfarlane and others would go on to exploit. If Mabey has been grouped recently amongst the 'new nature writers', he is also their precursor. As Matless notes, he 'has illuminated British cultures of nature since the 1970s', and his 'achievements and experiences encompass much of recent British environmental culture' ('Nature Voices' 179). Mabey's writing then and since has been crucial to this shift in two respects: it offered a revisionary account of what nature is and where it might be found, and it questioned the suitability of various narrative forms for best capturing this revised sense of nature. Though Mabey works with natural history traditions, his writing tends towards more personal than scientific reflection and forges innovations that have been taken up in fictional discourses, as well as imparting the benefits of the latter to non-fictional genres. *The Unofficial*

*Countryside* (1973) is the first in a series of Mabey's texts to question the place of nature: 'looking for wildlife we turn automatically towards the official countryside, towards the great set pieces of forest and moor. If the truth is told, the needs of the natural world are more prosaic than this. A crack in the pavement is all that a plant needs to put down roots' (12). With a change in focus and scale, the unofficial countryside – 'covering everything from a planned suburban playground to the accidentally green corner of a city-centre parking lot' – might provide an equally rich sampling of nature (14). Mabey's point is similar to that made later by Worpole. The shift is away from an aesthetic that brackets nature off from everyday life and towards one that 'confront[s] the modernised countryside' and assesses the relationship between the natural and the man-made (*Unofficial Countryside* 23). Due to the kind of landscapes that he visits, Mabey is concerned with locales that bear the marks of human intervention – canals, working and redundant quarries, car dumps – and with 'their wild inhabitants' (14). The new sites that Mabey envisages and that many writers have more recently followed him to are the marginal, the liminal, the edgeland.[4] The recognition of the braiding of the domestic and the wild that occurs in these places is one of the most important features of the rejuveated environmental imagination in a wide variety of texts.

For Mabey, change in location also necessitates a formal revision. He had originally intended to write *The Unofficial Countryside* 'as a travelogue' tracing journeys around London's perimeter, believing the account of walking to be 'a neat device for weaving … thoughts and observations into a narrative'. However, 'a programmed hike like this, with highspots strung out along the route, made the whole enterprise feel too much like an old country walk'. Alternately, a less 'enterprising' engagement might possess more flexibility in its attention to an environment's meaning: 'It was a change in focus that was needed, a new perspective on the everyday' (21, 24, 26). The form that the text takes as a result is one that resists the teleological pull of the quest narrative and is, instead, episodic in structure. The same attentiveness to narrative formations and deformations is the abiding characteristic of Mabey's work since, and (alongside the scoping out of edgelands) is its mostly widely felt legacy. *Beechcombings* (2007) is 'a brief history of the narratives we've constructed about trees over the past thousand years to make them accessible, useful, comprehensible and obedient'; but simultaneously attests to the ways in which the beech 'subverts clichés' and that '[b]eechwoods have occult lives of their own, in layers and time-frames we're not used to exploring' (xi–xii, 18). The text contests anthropocentrism in suggesting that '[t]he entire history of our relationships with trees could be seen as [a] kind of debate. We argue them into forms that suit us, they respond with tortuous narratives of their own' (26). More recently still, *Turned Out Nice Again* (2013) investigates 'that ceaseless, nagging narrative we British have about the weather',

[4] Mabey notes elsewhere that the ecological term for such a place is an 'ecotone': 'a zone where one habitat merges with another, creating something with a character more than the sum of its parts' (*A Good Parcel of English Soil* 7–8).

finding it to be both 'an incontestable feature of the physical world' and 'a creature of our imaginations' (6, 75).

Questions of observation, of description, and of the awarding of priority to certain kinds of narratives of place are thus raised by this intervention. *The Unofficial Countryside* embodies a genuinely haphazard wandering which, for Mabey, matches the environment traversed. The 'new perspective on the everyday' that he calls for, and, I will argue, that Macfarlane answers, requires consideration anew of modes of engaging with place and the means of producing its meaning that will result from this. Crucially, this shift in perspective also requires attentiveness to the variation of narrative forms that might be employed to suture together experience and representation of nature. Each of Macfarlane's texts, *The Wild Places* in particular, is characterized by a narrative arc that dramatizes a trajectory away from one mode of conceiving of place's meaning and towards another. They reproduce this change in the form and structure of the text, employing novelistic techniques of deferral and revelation. Thus, the first half of *The Wild Places* records journeys to the far-flung north and west of Britain and Ireland before the second half returns to the south and east – 'the undiscovered country of the nearby' – to find wildness in the midst of human settlement (225). *Mountains of the Mind* and *The Old Ways* embody comparable movements.

## Locating Wildness

Two recently published texts that fall under a broad nature or place writing umbrella exemplify the extent to which narratological concerns now pervade this literature. Madeleine Bunting's *The Plot: A Biography of an English Acre* (2009) is explicit in its exploitation of the slippage between plot-as-place and plot-as-narrative (as well as its debt to the life writing genre) to emphasize the revised idea of place that it tracks. Bunting's 'personal investigation' into the resonances for her own family history of a particular acre of ground 'sit[s] within a bigger story of how others have also used this bit of land and the relationships across time that I have with them' (6). Similarly, Jean Sprackland's *Strands: A Year of Discoveries on the Beach* (2012) employs titular double meaning (strand as beach but also strands of narrative) to signal the text's mimicry of Ainsdale Sands' indeterminacy and mutability as a place. 'The tideline', Sprackland writes, 'is an open book in a babble of different languages' (196). Both texts are clearly indebted to and draw on Macfarlane's work in this area. If, as Stephen Daniels and Hayden Lorimer aver, the 'new, consciously creative conjunction of personal biography and natural history' is 'in fact a renewed relationship that demands some critical reflection on its own history as a literary genre' (4), then Macfarlane's texts have been at the centre of this reflective endeavour.

This reflection, finding new narratives for understanding the modern wild, takes place most forcefully in *The Wild Places*, the second book of Macfarlane's trilogy, but *Mountains of the Mind*, the first, puts much of the groundwork in place. Here, the cultural forces that go into the construction of inherited ideas of wildness are discussed at length, before the later book recalibrates them wholesale. In *Mountains of the Mind*, Macfarlane occupies a kind of double position of narration. For much of the book he is enthralled by big mountains – in many ways, the old wild – whilst at the same time he deconstructs the ways in which they have been seen and conceptualized. He buys into these ideas, but also unpacks them. In 'Possession', the candidly titled opening chapter, Macfarlane describes being 'attracted by the bleakness [and] parsimony of the landscapes of mountain and pole, with their austere, Manichean colour scheme of black and white' (5). Like the two texts that follow it, *Mountains of the Mind* interleaves a history of ideas with personal accounts of Macfarlane's own trips to (that is, physical engagements with) the landscapes he discusses. The rationale that informs this structure is not merely one of variety for ease of reading; rather, shuttling back and forth between sensory experience and cultural inheritance mimics the book's central thesis. This is 'reading' landscapes, in which both processes inform the other:

> We *read* landscapes, in other words, we interpret their forms in light of our own experience and memory, and that of our shared cultural memory. Although people have traditionally gone into wild places in some way to escape culture or convention, they have in fact perceived that wilderness, as just about everything is perceived, through a filter of associations. (18)

Thus, a mountain is 'a collaboration of the physical forms of the world with the imagination of humans' (19). The text is careful to maintain this mutuality. Each subsequent chapter 'trace[s] a genealogy for a different way of feeling about mountains' (21). The text covers seminal moments in landscape history – Charles Lyell's *Principles of Geology* (1830), John Ruskin's *Of Mountain Beauty* (1856) – alongside other less well-known texts, charting important developments in both pre- and post-Romantic eras. It includes findings from artistic and scientific fields (including exploration), and, in keeping with the collaborative thesis, shows how they have interacted. Encompassing such genealogies, the book attempts to account for 'how a mountain can come to "possess" a human being so utterly' (20). If Macfarlane does not, in this text, overcome this possession, he certainly draws attention to the cultural (and therefore, to some degree, relative) grounds upon which his enthrallment rests. *Mountains of the Mind* surveys the ideas that *The Wild Places* comes to revise.

That said, the short final chapter of the earlier text does begin to gesture towards or, at the very least, make space for the changes to follow. Mountains, Macfarlane concludes, 'quicken our sense of wonder', they 'make us ready to credit marvels'. Being in these landscapes and experiencing them with our senses – given examples include swirls of water, pelts of moss, the touch of a snowflake – 'reignites our astonishment at the simplest transactions of the physical world' (275). In short, they condition our receptivity in a way that can, crucially, be taken to other more prosaic places: mountains 'return us to the priceless capacity for wonder … and they urge us to apply that wonder to our own everyday lives' (276). This is exactly the task that *The Wild Places* performs at length. Furthermore, the deferral of this revelation until the first text's end also foreshadows the arc of the second text. As such, *Mountains of the Mind* functions something like a prelude to *The Wild Places*.

Like the earlier text, *The Wild Places* also brings into focus the collaboration between imagination, cultural memory, and landform. It takes up the same approach to 'wildness' and, like the texts of Mabey, is structured around a shift in that idea. The text is concerned with the lexicon and grammar of space and, as such, is paradigmatic of a redefinition of terms that has been fundamental to new directions taken in nature writing: the notion of 'wild' has undergone a shift away from a landscape untouched by human intervention, towards one in which human and non-human influences enter into complex, fluid interactions. Contiguous with this modification, wild places are no longer seen as spaces without history, but conversely as sites in which histories remain traceable, even if partly overwritten by subsequent development. Scale has also undergone revision: writers now attempt to situate wilderness within networks of influence emanating from beyond its borders, whilst also accounting for their own situated perspectives. These features reveal the dialogues with North American traditions of nature writing to which various critics (Stenning and Gifford, above) have drawn attention. Gary Snyder, who has been at the forefront of the American tradition in recent decades, has argued that though the habitual idea of the 'pristine' wild as 'landforms [that] are entirely the result of nonhuman forces' persists, a more nuanced sensitivity is desirable: 'shifting scales, it is everywhere' (10, 15). From this basis, Snyder finds that 'a radical environmentalist position is in no way anti-human' (193). Indeed, if recent trends have emphasized this point, it has been part of the American tradition as far back as its nineteenth-century founding with Henry David Thoreau, who noted that 'it is in vain to dream of a wildness distant from ourselves. There is none such. … I shall never find in the wilds of Labrador any greater wildness than in some recess of Concord, i.e. than I import to it' (*Journal*, August 30, 1856; qtd. in Schama vii). Macfarlane renews this acknowledgement and the attentiveness to the near-at-hand that springs from it, deploying both in his writing on contemporary Britain and Ireland.

*The Wild Places* dramatizes the shift. It is organized around a series of journeys from Macfarlane's home in Cambridge, initially to locations conforming to an opening perception of a wild place as 'somewhere boreal, wintry, vast, isolated, elemental, demanding of the traveller in its asperities'. From this position, to reach a wild place is 'to step outside human history' (7). The subsequent narrative arc of the text revises this position almost completely, ending with a return to Cambridge. Its geographical scope is far more limited than that of *Mountains of the Mind*, which gravitates most often to the Scottish Highlands and the European Alps but also encompasses trips worldwide (the Canadian Rockies, the Nepalese Himalayas, the Tian Shan range in China). *The Old Ways* repeats the structure of *The Wild Places*: though part three, 'Roaming', takes in international trips, it is superseded by a return to England in 'Homing', the final part (*Old Ways* 207, 287). For Matless, 'the narrative trajectory' of *The Wild Places*, 'from Cambridge to relative extremities and back again, replicates longstanding senses of British cultural core and periphery' ('Nature Voices' 180). Kathleen Jamie also takes issue with this aspect of the text. In her reckoning, 'strikes into Scotland or Ireland or Wales are just that – strikes, then retreats', with the effect that 'Cambridge is still the centre of the world'. Indeed, her critique finds that Macfarlane's narratorial position, that of a 'lone enraptured male', reduces 'the variety of our engagement, our way of seeing, our languages' for landscape ('Lone Enraptured Male' 25–7). More charitably, Smith finds that the text 'is modelled on the form of the *bildungsroman* in which our narrator undergoes a coming of age or wisdom, an epiphany which, in this case, opens his eyes to the more delicate intricacies of nature and culture closer to home' (7). The texts embody a metropolitan perspective but simultaneously scrutinize the assumptions that this involves. They are more democratic in their marked subjectivity than Jamie perceives them to be.

Following the opening gambit of the wild 'outside human history', the initial trips are to the northern and western extremes of Britain and Ireland where the old idea of wildness is still partially realized. However, even here the traces of history begin to impinge on the concept of pristine wildness. At Strathnaver, in Sutherland on Scotland's north coast, the presence of ruins in the landscape provide evidence of former inhabitation before the highland clearances. In an inversion of the previous collocation, remoteness does not demarcate the absence of human history, but rather is produced by that history: '[l]ike so much of Scotland's wildest land, this is not an empty landscape but an emptied one' (51). Furthermore, this emptiness is a recent phenomenon following a long history of habitation: 'the first settlements had been established in Strathnaver more than six thousand years previously, and ... there had been human presence of some kind here more or less ever since. Marks of these successive occupations were to be found everywhere in the strath' (122). Macfarlane's understanding of place is dependent on an awareness of its longer historical narratives. '[T]he vision of wildness with which I had begun my journeys', Macfarlane goes on to note, 'was starting to crumble from contact with the ground itself' (127). The agency awarded to place in formulating the conceptual shift in this phrase is subsequently emphasized further: 'the river and the land seemed to caution against categorical thinking, against partitions' (127). As is the case in *Mountains of the Mind*, the land itself and our imagining of it combine to produce a new way of seeing.

The revisionary narrative that the text as a whole performs is structured around a central contrasting pair of encounters, roughly at the text's centre point. The first, in the chapter entitled 'Summit', is at the book's apex – its highest point and furthest north (conceptually, if not quite geographically). From here, the arc of the text descends, moves south, and shifts register. The summit of Ben Hope is the landscape that most readily conforms to the initial idea of wildness, containing '[n]othing human, only the mineral marks of fossils and glaciers' (157). However, whilst there Macfarlane 'felt no companionship with the land, no epiphany of relation', and concludes (with a telling phrase that constructs agency differently than in the foregoing example) that '[t]his place refused any imputation of meaning' (157). The Romantic image of the poet or artist communing with pure nature is refuted here. Similarly instructive terminology appears in Macfarlane's subsequent work: *The Old Ways* considers that a landscape's 'refusal of relation might offer its own bracing reward' (339), and elsewhere Macfarlane suggests that a place's 'unfetchability' might be 'intellectually conducive' ('Way-rights' 15). In contrast to the absence of history and connection at Ben Hope, the following chapter of *The Wild Places* details a journey to the Burren, where the text's central revelation occurs. This area of limestone pavement on Ireland's west coast bears the marks of both the geological and the human forces that have produced it. Macfarlane and Roger Deakin, who is accompanying him on this trip, arrive at a large gryke (a fissure in the limestone pavement):

> We lay belly-down on the limestone and peered over its edge. And found ourselves looking into a jungle. Tiny groves of ferns, mosses and flowers were there in the crevasse – hundreds of plants, just in the few yards we could see, thriving in the shelter of the gryke: cranesbills, plantains, avens, ferns, many more I could not identify, growing opportunistically in the wind-blown soil. The plants thronged every available niche, embracing one another into indistinguishability. Even on this winter day, the sense of life was immense. (168)

The contrast with the experience on Ben Hope is made clear: 'This Roger suddenly said … is a wild place. It is as beautiful and complex, perhaps more so, than any glen or bay or peak. Miniature, yes, but fabulously wild' (168). The shift that results is one of scale and of time-scheme. Macfarlane's application of the terminology of the vast to the small scale emphasizes the change: he and Deakin peer over the edge of the crevasse as though it were a cliff, and the plant life to be found there constitutes a jungle. If a revision occurs here, certain inherited facets of encountering a wild or natural place are also retained. The engagement that this passage describes is, like most Romantic landscape writing, scopic in character. It also retains the elevated prospect perspective that surveys the scene below (even if this is on a much smaller scale). Furthermore, elevated viewpoints characterize the tree-climbing passages that open and close *The Wild Places*, though other more embedded perspectives do receive extended treatment in other passages (see below). The break from tradition, then, is not one of complete severance, but of selective revision. Older practices do have an ongoing role to play in the Macfarlane's nature writing, if in recalibrated form.

The lexicographical technique employed in this passage describing the gryke – the mixing of the large-scale terminology and small-scale landscape feature – informs the rest of the text after this revelatory moment. In 'Storm-beach', Macfarlane visits Blakeney spit, where he 'lay down on the warm dry shingle, and sighted the spit's summit ridge off ahead of me. I looked along it at the pebble level' (243). Features associated with mountain terrain ('summit ridge') are suited to the low-level, gentle gradients of a coastal landscape when perspective is shifted to the 'pebble level'. The supine position adopted here is in many ways the opposite of the elevated prospect before. In this instance, and at the gryke with Deakin, it is clear that a reconceived wild is predicated on a renewed attentiveness to the overlooked. Macfarlane's text foregrounds Deakin in recognizing the need for attention to the small scale, but Mabey is also a precursor. *The Unofficial Countryside* records how 'the conditioned urban eye' habitually registers only the 'big set-pieces', whilst smaller plants and other instances of wildness 'need the slow look, the close-up, ways of seeing that we are not used to in places where the snatched side-long glance is the convention' (34).[5]

For Macfarlane, whilst 'the old sense of the wild' is wedded to 'an ideal of tutelary harshness', the gryke exhibits 'another wildness at work: an exuberant vegetable life, lusty, chaotic and vigorous. ... The wildness of the gryke ... was to do with nowness, with process. It existed in a constant and fecund present' (176–7). The reconsidered time-scheme in which the new wild exists might also be related to Mabey's work. In *Beechcombings*, it is noted that the lives of trees are 'governed by rhythms alien to us' – 'whole civilisations may come and go in the lifetime of a single veteran beech' (29–30). *Mountains of the Mind* includes similar revelations: appreciating geological deep time, mountains can no longer embody 'effigies of permanence', but acquire 'an exciting, baffling mutability'; 'through the spectacles of geology, *terra firma* becomes *terra mobilis*' (35, 43). The extension of frames of reference in this passage and the emphasis on the 'fecund present' of the gryke share a common root in the reconsideration of the time scales imposed on place and nature from habitual human perspectives. After the Burren episode of *The Wild Places*, the journeys that the text describes forego the furthest reaches of the archipelago and return to the familiar and seemingly un-wild locations of the south and east, where the new understanding of the wild is 'not as something which was hived off from human life, but which existed unexpectedly around and within it' (226). In another register, the shift envisaged over this narrative can be conceived geologically: 'in my journeys, I had moved down a gradient of mineral resilience – from the igneous coastline of Coruisk and the volcanic north-west, through the soluble limestone of the Burren, down to the sandstone and chalk of the Holloway counties, and now to these transient and fraying south-eastern edgelands' (242). This 'mineral dissolve' is complete when Macfarlane's text reaches the tidal mud-flats of 'unlikely wild Essex' (272, 315). Again, the text is concerned to foreground this trajectory in the shift between, and the layering of, variant registers for the description of place: Macfarlane's inclusion of the geological here, alongside numerous other ways of thinking about landscape, works to support the text's 'polyphonic' credentials.

After the conceptual shift at the text's apex, subsequent journeys seek out locations that express a braiding of human and non-human forces that produce places characterized by wildness. The visit to the holloways of Dorset illustrates

---

[5] Mabey's reconsideration of scale in this text is also suggested by his terminology: a small patch of neglected ground is considered as a 'wilderness' (a term usually reserved for large-scale vistas), whilst Hampstead Heath is called an 'expanse of wild country' (*Unofficial Countryside* 44, 106).

the point most emphatically. In an echo of the remoteness of the Scottish highlands that resulted from clearances recorded earlier in the text, these spaces of wildness in the midst of a thoroughly managed and cultivated landscape have been produced by and would not exist without a history of human intervention and land use. A holloway, Macfarlane notes, is '[a] route that centuries of use has eroded down into the bedrock, so that it is recessed beneath the level of the surrounding landscape'. As such, they constitute 'landmarks that speak of habit rather than suddenness. ... They are the consequence of tradition, of repeated action' (216). It is this habitual use, followed by abandonment, that has led to their re-wilding:

> Few holloways are in use now: they are too narrow and too slow to suit modern travel. But they are also too deep to be filled in and farmed over. So it is that, set about by some of the most intensively farmed countryside in the world, the holloways have come to constitute a sunken labyrinth of wildness in the heart of arable England. (218)

The components of the new conception of wildness come together in this experience. This wildness does not represent an escape from broader networks of human history and land use but is produced by them – it is a fissure in the midst of intensive agriculture. A revaluation of the scale of wildness is presented. The landform itself, being a hollow, precludes the panoramic vista from mountain-top that is associated with the old wild. Here, the hedges on each side 'had become linear forests, leaning into one another and meshing above the old sunken road to form an interlocking canopy or roof, turning road into tunnel' (223). If the episode at the gryke retained an elevated perspective in some recalibrated form, here it is abandoned altogether. In these gradual changes the narrative arc of the text is one of incremental modulation rather than a sharp break. The resulting perspective is one that views nature not from outside but from within. In *Holloway* (2012), Macfarlane's subsequent collaborative piece revisiting this place, the same strong demarcation of inside/outside is made: 'dropping down by the old ash-tree into the near darkness' of the Holloway, 'the bright hot surface world was forgotten' (12–13). If the elevated perspective has been exchanged for the supine at Blakeney, here the subterranean is adopted. In *The Wild Places*, it is the terms of this description – characterizing hedges as 'linear forests' – that serve to blur the distinction between the familiar and the exotic, making strange that with which one is supposedly familiar.

*Holloway* offers the grounds on which to make further points here. As another narrative of the same place on a different visit, it serves as an intertext for the account in *The Wild Places* and, in its differences, emphasizes that what is being recorded is an experience of place in time. Forgoing the explanatory passages that the latter text includes, *Holloway* is more consciously artistic and highly wrought. Macfarlane collaborates with artist Stanley Donwood and writer Dan Richards to produce a book that contains multiple renderings of this place and this trip (two textual, one pictorial) that, taken alongside one another, emphasize individuated experience. No explicit designation is given to signal where the section written by Macfarlane ends and that by Richards begins, but it is nonetheless abundantly clear from the stylistic differentiation between the two. This sharp distinction also serves to highlight textuality and the idiosyncrasies of each version. Counter to Jamie's assertion, and perhaps in response to it, this text looks to expand the variety of engagement, of ways of seeing, and of languages for landscape. Indeed, in interview, Macfarlane has spoken of the danger of writing for landscape edging towards a 'solipsistic grotesquerie' and pondered 'how to enable the … ecologically liberal instinct towards acknowledging other points of view and valuings, without deliquescing into a gloop of equalisation' (qtd. in Stenning 81). *Holloway* embodies a considered response to this danger: it is democratic, yet informed; open, but not unduly levelling. The same commitment informs Macfarlane's fascination with paths in *The Old Ways*: they are interesting, at least in part, because of the democratic sense of place that their 'consensual making' by many different people over a long period of time exhibits, and because they are 'collectively made for general use', thus bespeaking community (17). In the same text, Macfarlane writes that intimate knowledge of a place does not lead to 'comprehensive totality', knowing it utterly; but rather that 'long acquaintance' teaches that 'familiarity with a place will not lead to absolute knowledge but only to further enquiry' (111). Macfarlane's texts emphasize his personal experiences of place (and his preparatory reading about those places) not to bolster his authority, but to underline his subjectivity (or, in novelistic terms, his fallibility as a narrator).

## Writing the Wild

Like Sinclair in Chapter 3 of this study, Macfarlane's representation of place and nature deploys techniques of defamiliarization in relation to the quotidian and the everyday that can be traced back to Shklovsky's *ostranenie*. This effect registers in both the descriptive aesthetic and the shape of Macfarlane's prose style. Descriptive techniques such as blending the familiar and the exotic, and applying the vocabulary of the vast to the small-scale landscape feature, are repeated throughout the latter half of *The Wild Places*: '[l]ying just off the Suffolk coast is a desert' at Orford Ness (241), 'the Essex landscape' is 'as inspiring and elemental as the Pamirs or the Arctic' (274). A similar method informs the opening chapter of *The Old Ways*. Here, Macfarlane describes a winter evening walk amidst the prosaic edges of Cambridge. Though he takes a very familiar route – 'the path I've probably walked more often than any other in my life'– it is made strange by the conditions: 'the snow caused everything to exceed itself and the moonlight caused everything to double itself' (6). Following another trail, he comes upon 'a surreal landscape' of swooping expanses interspersed with 'big white humps [that] surrounded what appeared to be a small neat lake' (7). The description preserves uncertainty ('what appeared to be') and initially withholds the crucial locating detail: 'edited of its golfers by the darkness, transformed by snowfall and moonlight, the county's most exclusive golf course had become a strange realm of open country' (8). One of the archetypal features of suburban countryside is indeed made strange in these terms.

If descriptive passages bear the mark of defamiliarization, so too do the books' larger structures. Macfarlane's readings of others' texts often draw attention to the way in which the form of a book might 'resemble the landscape in which it had been conceived' ('Way-rights' 18). The same connection is evident in his own work. He has spoken of his ambitions for *The Old Ways* as a book 'proceeding by pattern rather than plot', and that 'formally or technically speaking' it would exhibit 'high reflexivity combined with assiduous pattern making'. The rationale for this ambition is 'to make language and form perform some of the recapitulations and overlaps I had experienced while walking these paths' (qtd. in Stenning 78). The mode of representation is fitted to a recuperation of the experience that it describes. Thus, if pathways are taken to be conduits of the past – 'rifts within which time might exist as pure surface, prone to weird morphologies, uncanny origami' – then the book's patterned rather than strictly plotted construction mimics this defamiliarizing characteristic with the aim of 'bringing [about] alignments and discrepancies, elective affinities, shifts from familiar dispositions' (*Old Ways* 22).

Macfarlane's editorial comments elsewhere serve to describe the investments that his own writing is making here. Writing about natural history filmmaking of the 1920s and 1930s, Macfarlane has noted affinities with Shklovskian modernist forms, and that both aim to 'enhance perception of the familiar, to de-automatize the quotidian'. If Shklovsky described techniques for making the stone stony again, Macfarlane continues, 'BIF films, with their new filmic techniques of time magnification and altered motion, were also in their way devoted to defamiliarising the commonplace, to making the flower once again flowery' ('Very ordinary, very beautiful' 17). Furthermore, in his introduction to the 2005 reissue of J.A. Baker's *The Peregrine*, an important book for him as for numerous other contemporary nature writers, he suggests that it is the 'high-blown stylistic artifice' that generates this text's importance (xii). Again making reference to modernist precedents, Macfarlane finds that Baker's 'stunning set-piece descriptions' are 'as formal and dynamic as any Imagist poem', with the effect that prosaic Essex is made 'wonderfully unfamiliar' and imbued with a 'quality of deep strangeness' (xii, xiv). Macfarlane's own texts inherit this tradition: a critical awareness certainly feeds into *The Wild Places* and the 'undiscovered country of the nearby' that it comes to take as its subject matter (225). Wildness, in short, exists not just in remote corners, but also as an eruption, or perhaps an interruption, in 'the margins, interzones and rough cusps of the country: quarry rim, derelict factory and motorway verge'. This version of wildness is only available from an altered perspective, refracting the local and familiar, that Macfarlane describes as 'myopia of a good sort' (227). The 'wildness as a state of mind' advocated in his foreword to *A Wilder Vein* is thus achieved through a kind of modernist-indebted wild writing (vii).

That foreword also identifies another of the salient features of Macfarlane's nature writing: 'cognition is site-specific, or motion-sensitive: … we think differently in different landscapes' (ix; see also *Old Ways* 27, 'Way-rights' 13). A similar point is made frequently in *The Wild Places*: 'Time is kept and curated in different ways by trees, and so it is experienced in different ways when one is among them' (100); 'I had … learned in each place [visited], had been bought to think by each in unexpected accents and shapes' (115); '[t]o be out by night in a forest, by a river, on a moor, in a field, or even in a city garden, is to know it differently' (202). In *The Old Ways*, Macfarlane develops terms of description that capture site-specific cognition. Thus, a typical passage describing the flow of thoughts in a particular place reads 'We sat on the out slope of the sea wall, easting sandwiches and talking. … I thought of the curlew I had seen earlier that year, and about how the paths of birds and animals were really the oldest ways of them all'; or, 'Sitting there in the sun, looking up at Minya Konka, I thought I had undergone a similar conversion' (17, 280). The thought of bird migrations (in the first example) or a conversion (in the second) is not all that is reported. Rather, the thought is caught up in clauses that locate the process of thinking in time and place. Its site-specificity is presented. Daniels and Lorimer find 'a deeply personalized quality of expression' to be the 'signature feature' of new nature writers like Macfarlane (4). This feature is not necessarily as self-aggrandizing as some have found it to be. In these passages, it appears as a means by which to recuperate experience in textual form. In his emphasis on the necessity for experiencing place first-hand, Macfarlane is entering into debates surrounding the ways in which landscape is apprehended, and drawing on contrasting critical paradigms for the study of place ranging from landscape as realm of embodied practice, to landscape as symbolic text. He is, in short, attempting to find ways to consolidate the two.

The journeys that make up *The Wild Places* retrace a wide range of textual antecedents and are clearly preceded by a large amount of preparatory research and reading that Macfarlane regularly refers to. Whilst there is undoubtedly a reliance on prior representations in the apprehension of place here, it is accompanied by a thorough-going commitment to enter into sensory engagement with and experience of the environments encountered. Furthermore, Macfarlane is at pains to stress a tactile and aural engagement alongside a visual one: the opening passage of the text, describing a beech wood, combines scopic, aural, and haptic modes of encountering, and thus sets the tone for the rest of the book. The passage comprises three paragraphs devoted, in turn, to the sounds – 'the noise of the wood in the wind' – the sights – 'sunlight [falling] in bright sprees on the floor' – and the feel – the trees' 'sagged and wrinkled' bark – of the wood (3). The text as a whole is informed by the conviction, drawn from phenomenology, that 'our minds are shaped by the bodily experience of being in the world', and that '[a] constant and formidably defining exchange occurs between the physical forms of the world around us, and the cast of our inner world of imagination' (203). *The Old Ways* also proceeds in kind. Dialogue between the outer and the inner is predicated on particular forms of engagement and result in a sense of wonderment. The text is structured to reflect this: passages containing historical accounts of place and other scholarly material are intersected by descriptions of walking, swimming, climbing, and sleeping in those same places. This structure illustrates the entwining of the two modes of knowledge (representational and experiential) that combine to produce an idea of wildness. Indeed, the two strands are inseparable in that experiences of place underpin representations thereof, which, in turn, condition experience. The idea of the 'deepwood', Macfarlane notes, 'flourishes in our architecture, art and above all in our literature' as a place of myth and dream, but 'there is no mystery in this association of woods and otherworlds, for as anyone who has walked in woods knows, they are places of correspondence, of call and answer' (*Wild Places* 98).

In this exchange, some forms of representation are found to be better at capturing experience of place than others. In a discussion of different types of mapping, Macfarlane contrasts a modern sense of cartography – 'an endeavour of exacting precision, whose ambition is the elimination of subjectivity from the representation of a given place' – against an earlier version: 'In its pre-modern expressions, mapmaking was a pursuit that mingled knowledge and supposition, that told stories about places, that admitted fear, love, memory and amazement into its projections.' 'Story maps', he goes on, 'are organised around the passage of the traveller, and their perimeters are the perimeters of the sight or experience of that traveller' (*Wild Places* 141). It is clear that Macfarlane conceives of his text as a map allied to this pre-modern mode of 'narration as navigation' that seeks out rather than excludes wonder (144). In one of the more polemic passages of the text, the distinction gravitates around this term: the 'neutral organisation of data' that characterizes the road-map 'encourage[s] the elimination of wonder from our relationship with the world. And once wonder has been chased from our thinking about the land, then we are lost' (145). For Macfarlane, wonderment may be a 'projection', but it is one that arises out of 'the perimeters of experience'. The composite form of *The Wild Places*, assimilating different modes of understanding into a narrative of place, is coupled with a partiality that is perhaps the defining perspectival register of his nature writing and the poly-generic literary non-fiction form that it takes.

Whilst a more explicit politics of engagement is present only in passing, then, wonderment is a crucial component of what Macfarlane's texts call for and the action that they intend to inculcate. Simon Cooke has argued that the renewed sense of wonder that characterizes late twentieth-century travel narratives 'is not simply an escape from the sources of modern disenchantment, but a profound response to them' (2). This is certainly the case with Macfarlane. By way of a coda to the chapter, I wish to suggest that Macfarlane's writing possesses a thoroughgoing environmental politics, even if this remains, for the most part, only implicitly articulated. Intimate knowing, in the form of restoring and maintaining a relationship with and a feeling for place, is political in forming a first step towards acting differently. If more openly campaigning forms of environmentally oriented literature risk being pigeonholed by those who are not already converted to the cause, then the subtler positions adopted here preserve integrity and are less easily dismissed. Participation in ecology is also part of this politics. In his introduction to Richard Jefferies's *Nature Near London* (1883), Macfarlane argues that Jefferies's 'writing worked at what might now be called a consciousness-raising exercise', and that he 'sought to show his readers their participation in a continuum of human and non-human existence' (xx). The same might be said for his own work. This commitment informs the choice of landscape– 'the undiscovered country of the nearby' in *The Wild Places*, paths as the conduits of people's experiential relationship to place in *The Old Ways* – that Macfarlane's texts gravitate towards. In this way, his writing responds to Worpole's assertion that 'it is vital that ... unassimilated landscapes ... are documented and valued if they are not to be levelled or "improved" in the name of some larger political programme'. Thus, they are part of the 're-inscription of the landscape', the 'new counter-reformation' that 'is underway' (*New English Landscape* 73).

# Chapter 5
# Noticing: Kathleen Jamie

The title poem of Kathleen Jamie's *The Overhaul* opens with a simple exhortation to 'Look' (28). *Sightlines*, her most recent collection of essays, suggest that 'careful recording' is 'a wholly different way of looking' (156). This chapter enters into a comparative relationship with that which immediately precedes it: Jamie's writing offers differently accented refractions of a similar set of concerns to those of Robert Macfarlane, maintaining a more sustained critique of habitual ways of seeing. If Macfarlane's texts are testimonial in their approach to nature and place, Jamie adopts a more documentary tone focused on attentive recording in her prose texts, *Findings* (2005) and *Sightlines* (2012), and in her poetry, particularly her most recent collections, *The Tree House* (2004) and *The Overhaul* (2012). Rather than seek a sense of narrative coherence, Jamie's prose foregrounds the place where patterns do not hold. The characteristic spareness of her poetry fulfils a similar function. Jamie's poetics – in prose and verse – create an aesthetic of reticence in the description of nature. On a larger scale, her episodic texts consciously resist the temptation to formulate grander narratives. Notwithstanding this humility, this chapter explores the ways in which Jamie attaches a reverential aspect to noticing. Her writing bears out her commitment that language is not that which gets in the way, but what connects us with the world.

The critical comparison of Macfarlane and Jamie is not new, but the account offered here does not concur with the stark contrast depicted in most appraisals. Rather, I work from the distinctions others have observed towards a sense of the areas in which their writings are also convergent. If Macfarlane is a central figure of the 'new nature writing', Jamie is often cast as a detractor from the consensus. Two ways in which critics have articulated this difference are instructive in outlining my own argument for this chapter. They concern perspective and tone. Matless takes issue with the 'conspicuously wrought prose' that he finds in Macfarlane's writing, and suggests that 'those seeking a contrast' might read Jamie's *Findings*, published two years earlier, 'where authorial presence in Scottish landscape is made via a carefully honed colloquial modesty' ('Nature Voices' 186). A national divide underlies the variance between the two writers' texts in this account (and others), by which Macfarlane's English perspective is contrasted in Jamie's Scottish evaluation. Jamie's credentials as a Scottish writer interested in Scottish issues and identity is well established from her earliest poetry in *Black Spiders* (1982) through *The Queen of Sheba* (1994) and *Jizzen* (1999) and into the two collections discussed here, as well as her editorial work for *New Writing Scotland* in the 1990s. Her use of a colloquial Scots English for poems in all of these collections emphasizes a distinct identity that is most recently underlined in the translations of several of Hölderlin's poems into Scots in *The Tree House* and *The Overhaul* (cf. *Tree House* 12, 28; *Overhaul* 38, 48). It is not surprising then that she is alert to the politics of who writes about and for Scotland's places.

Scottishness certainly informed Jamie's hostile review of Macfarlane's *The Wild Places* (discussed in Chapter 4) and the cultural colonialism that she finds there. The comments that she has made in interview since that review have usefully clarified the point of contention: 'he came up here from Cambridge into the Celtic countries, Scotland and Ireland in particular, and went around saying "it's so wild". It's so not! ... There've been Mesolithic people, Neolithic people, Bronze Age people. These landscapes have been humanised for thousands of years. There's nothing wild in this country' (qtd. in Goring n.p.). The most obvious reading of this is as a territorial dispute in which Jamie resents a voice from the metropolitan south, from England, speaking for Scotland, casting her as the 'wild' other, and implicitly silencing her native ('Celtic') writers. However, I want to suggest that the argument is not just one of 'local places for local people' but a more nuanced advocacy (albeit in negative terms) for sustained ways of seeing that are only possible via the duration that habitation provides. Whilst Macfarlane is often a visitor to the places about which he writes, Jamie is more often a resident. This perspectival difference is one that feeds into a tonal difference. In *Sightlines*, Jamie makes a similar point, observing that nowhere [in Scotland] is truly wild or abandoned' and stating her distrust for 'any starry-eyed notions of the "wild" or the "remote"' (135, 143). The 'starry-eyed notions' induced by newly visited places contrasts the sceptical stance that Jamie's own writing adopts. The latter is born of long-standing familiarity. Thus, where Macfarlane's work is enraptured, Jamie's highlights 'attention'. As such, the balance between documentary and testimony in their texts signifies a key distinction (that informs the difference in 'authorial presence' that Matless notes). The difference is not absolute – Jamie is occasionally rapturous, just as Macfarlane is by turn measured – but a contrast in pervasive tone does inform the comparison drawn here.

Chapter 4 began by identifying the salient features of Macfarlane's place writing through a study of his essay, 'The Counter-Desecration Phrasebook', in *Towards Re-Enchantment* (2010), edited by Gareth Evans and Di Robson. A similar tactic is adopted here: the comparison between Macfarlane's aesthetic and Jamie's is aided by the fact that she also contributed an essay to the same collection. 'On Rona' is Jamie's contribution to *Towards Re-Enchantment*, a version of which also subsequently appears as one of the essays that make up *Sightlines*. The essay is simply titled, giving little more than the location visited and described, though the phrasing of the title also hints at the traditional construction for indicating an essay's topic (e.g. 'On Friendship', 'On Liberty', etc.). Thus, it is implied, the visit and the record of the visit remain in close proximity to one another – a point born out in Jamie's characteristically spare prose style. Jamie records a research trip of several days to the isolated and now-uninhabited Scottish island of Rona with archaeologist Jill Harden and ornithologist Stuart Murray. Harden is investigating the ruins of former inhabitants' buildings (including those supposed to have been used by Saint Ronan during his time on the island), whilst Murray is recording numbers of storm petrels nesting there.

Travelling in the company of two specialists with particular knowledges allows Jamie to develop a different kind of engagement with place than that of Macfarlane. Jamie notes, in a repeated refrain, that '[t]hose two, Jill and Stuart, were great observers'. The focus on observation acts as a measure that places limits on wonder: 'Jill was an archaeologist and, like Stuart, not one to be fanciful' ('On Rona' 136). However, the absence of wonder is not absolute here. Rather, for Jamie, faith in seeing and observation is possible when conditioned by the necessity of accuracy: 'Stuart's of that sterling tradition of self-taught naturalists; a bird-man who says, "believe what you see", but a prerequisite of that believing is a great accuracy of seeing, and a rough idea of what you're looking at' (136). Wonder is not absent, but it is tempered. In addition to 'accuracy' in the moment of engagement with nature and place, this passage also demonstrates an awareness of what is brought to that engagement: Jamie expresses a respect for the skills of looking and observation built up over years of practice, and recognizes a multiplicity of ways of measuring or describing the land (the archaeologist's dig, the naturalist's survey, and the writer's account). *Findings* and *Sightlines*, as will be seen, focus on similar issues of observation and attention at length. These emphases indicate a different register of ideas of intimacy with place, and of modes of describing that intimacy. If intimacy here does not emanate from long familiarity (there are, in modern times, no inhabitants of Rona to fulfil this role), it is in part the result of specialized knowledge (archaeological and ornithological). It is also conditioned by a relationship of labour with the landscape. Jamie assists Murray in mapping the petrels' nests in holes across the island: 'It was a curious task, very intimate, to sail to a faraway island, then crawl over it on hands and knees, like pilgrims or penitents' (142). Intimacy here is one of physical contact through work, but Jamie's description likens this to an act of faith.

A pilgrimage motif runs throughout the piece, linking the current engagement with place to Rona's history as a site of religious retreat. Whilst the terms of description for contact with nature and place implicitly emphasize an awareness of this spiritual history – 'Stuart spoke to the birds, Jill communed with the stones' – Jamie is also careful to exercise restraint in the more explicit imposition of meaning by connection with the past (143). Whilst working with Jill in the abandoned settlement's graveyard, Jamie recognizes the appeal of speculating about former inhabitants: 'Of course it made us think of them. The long dead people whose graves we knelt on. We called them "them" and spoke about them every day.' Similarly, Jamie is also intrigued by Ronan's cell inside the ruins – 'I crept in just to wonder what he did in there, Ronan; to imagine him right there' (144, 146). This is a different kind of 'wonder' though – closer to inquiry than enthralment. Again, marvelling is tempered on this subject:

> Whether he was really alone, as romanticists would have it, or whether others came with him – monks, lay penitents, men without women, – well, as Jill would

> say, we don't know, do we? Surely it would have taken more than one to do the spadework; even saints must eat. And if there were people on Rona already, watching as the Christians' boat drew nearer – we don't know that either. (145)

Jamie is keen – here as elsewhere – to purge traces of romanticism from her writing of place (far more so than Macfarlane).[1] Any speculation that is undertaken here is based on a practical assessment of the labour required to live in this environment and the necessity of a group to do the 'spadework'. A comparable circumspection is central to much of Jamie's work. In an 'Author Statement' for the British Council, she makes a comparable point in more general terms: 'I couldn't even say what I write "about", because I distrust the relationship expressed by the word "about". I'd rather say I write "toward". Or perhaps "within". At the moment, I'm writing a lot "toward" the natural world' (n.p.).

The sharpest contrast to Macfarlane's writing, though, is structural. Jamie's question ('we don't know, do we?') is truly rhetorical in receiving no answer. Where Macfarlane's nature writing is, in part, based around a sense of narrative coherence, Jamie's foregrounds the limitations to which this patterning might aspire. This passage is unwilling to go further in its speculation and so ends pointedly unresolved – it distinctly marks its refusal to impose meaning. Jamie's piece on Rona does implicitly formulate links between the fluctuations in human inhabitation there and the bird migrations, but these patterns remain more implicit and less sharply accented than similar aspects of relationship to place in Macfarlane's work. Again, their two methods are not opposed to one another, but rather signify variant techniques for approaching a shared set of concerns. Alongside the comparative points outlined already, one of the chief divergences between the two writers is the way in which narrative structures inform Macfarlane and Jamie's different modes of engaging with nature.

Jamie's visit to Rona in this text, as well as the religious tropes and pilgrimage motifs through which that visit is described, are indicative of a pervasive set of images and practices shared with Macfarlane, with the other writers approached over the course of this book, and perhaps, for that matter, with much contemporary literature of place. Ideas of sacred space and holy places are leitmotifs running through all of these texts. The first journey of *Findings* is to Maes Howe, a Neolithic chambered cairn and tomb on Orkney 'where, 5000 years ago, they buried their dead' (10); Macfarlane's first destination in *The Wild Places* is Ynys Enlli (Bardsey Island), the 'island of 20,000 saints', where early Celtic Christians 'built monasteries, cells and oratories, dug cemeteries for their dead and raised

[1] It is a small 'r' romantic tendency rather than capital 'r' Romanticism that Jamie is purging from her own writing. Her more complex relationship to Romanticism is detailed later in this chapter. Suffice to say here that the critical reception of Jamie's work has formulated a multifaceted response to Romantic writing. Louisa Gairn has noted that '[w]hile Romanticism undoubtedly continued, and continues, to be influential, realisations of its limitations have provoked varied responses in the formation of … models of attentiveness, observation and representation' (4).

stone crosses to their God' (24). If both writers distance themselves from inclusion in these faiths ('they' not 'we'), their itineraries are defined by ancient pilgrimage routes and reliquary sites. Their imagery is also regularly refracted through a religious prism – such metaphors are at the heart of Macfarlane's '*polytheistic*' sense of 'wildness as a state of mind' (Foreword to *A Wilder Vein* vii). The same is true of the other writers here: Sebald's 'English pilgrimage' gravitates towards and loops together a series of ruins treated as holy sites for profound reflection, Carson's engagement with the religious and sectarian geography of Belfast is extensive, Sinclair's fascination with religious locations – former monasteries, churchyards – is accompanied by an awareness of the 'ritual purpose' of his walking. Recourse to these material landscapes and to this vocabulary for processing them has offered a useful mode for inflecting the *historical* value of place. All forms of pilgrimage – as detailed in Chapter 1 – involve some form of 'reference to the past' (Stopford x). Following often well-established routes has also allowed writers to play off recognizable traditions in their diverse versions of articulating place anew. The recuperative impetus that informs their work combines reiteration of textual sources and re-enactment of historical practices (that is, representational and experiential aspects) in the deployment of pilgrimage tropes.

## Attending to the Local

On the summit of Ben Hope, as discussed in Chapter 4, Macfarlane 'felt no companionship with the land, no epiphany of relation'; '[t]his place refused any imputation of meaning' (*Wild Places* 157). If this failure to connect with a place that readily conforms to received ideas of the wild provides an impetus towards reconceiving wildness, it might also be said that some of the inherited *structure* of engagement is retained here and through successive sections of *The Wild Places*. Again, questions of seeing, of ascribing or of refusing to ascribe significance, are at stake here. An 'epiphany of relation', another religious trope and often the source of a sense of wonder, is subsequently sought by Macfarlane in different places, but it remains the basis for connecting to a place.

By contrast, Jamie's *Findings* stages and undercuts exactly this kind of moment. In the first essay, 'Darkness and Light', visiting the Neolithic tomb at Maes Howe, Jamie reaches out 'instinctively' to touch the wall carvings to be found there 'to make a gesture of connection'. However, at this point the scene switches to a bathetic register when she is asked by the accompanying guide not to touch as '[t]oo many wandering, sweaty fingers would wear the carvings away' (18). The mode of engagement that seeks to make 'gestures of connection' or experience 'epiphanies of relation' is undermined. *Sightlines* adopts a similar approach. Its first chapter, 'Aurora' also works to distance the text from this kind of impulse early on. On a sea voyage to the Artic, one of Jamie's fellow passengers is given the role of Macfarlane-like figure searching for meaning in the icebergs that pass the ship: 'Someone calls, "they're so organic!" But organic is just what they're not. Their shapes and forms are without purpose, adapted to no end. They are huge and utterly meaningless' (6). Jamie's narrator is less deceived: 'the cameras click, but the icebergs give nothing, suggest nothing but a white nihilism'; 'Although they pass slowly and very close, I smell nothing but colossal, witless indifference' (7, 10). If Macfarlane stages arriving at this kind of revelation part way though his text, Jamie's accounts of the same recognition is less dramatized and comes much sooner in her books. In short, they get this baggage out of the way early in order to move on.

Jamie's critical review of *The Wild Places* finds that its 'lovely honeyed prose' acts 'like an enchantment on the land'. She assesses 'the association of literature, remoteness, wildness and spiritually uplifted men' to be found there, and relates it to 'a tradition' of such writing – structured around epiphanies and imputations – that 'remains largely uninterrogated' ('A Lone Enraptured Male' 26). Her hostility is, in part, informed by a concern that the continued dominance of this tradition is reducing 'the variety of our engagement' with landscape and place (26). Whilst the review does not always take account of the narrative arc of Macfarlane's book, taking the opening premises that it works to revise as a definitive rather than a provisional statement, it is indicative of the different nuances of engagement to be found in Jamie's own work.[2] *Findings* is preoccupied with questioning what she sees as the lingering romanticism in the 'imputation of meaning', and seeks to elucidate other forms of engagement, reconsidering Romantic legacies. As such, it recalibrates Macfarlane's wonder, and shifts the focus of nature writing towards issues of observing, noticing, and paying attention.[3]

In many ways, Jamie's texts start from the position that Macfarlane's dramatizes arriving at: here, the domestic and the wild are thoroughly entwined. For Jamie, the 'unfamiliar country of the nearby' is more locally focused than it is for Macfarlane. *The Wild Places* is structured around a series of journeys that work towards and subsequently descend from a highland apex: its trajectory pivots around the central revelation at the gryke on the Burren discussed in the previous chapter. *Findings*, on the other hand, interleaves journeys within Scotland with a more detailed and extended treatment of the locale in which Jamie lives (often finding wildness on foot or by bike from home), in the format of a series of related but discrete essays. *Sightlines* also focuses for the most part on Scotland, and where it does go beyond these borders it is to Scotland's sea-facing neighbours to draw out similarities. This difference informs Jamie's reception of *The Wild Places* –

---

2 In a subsequent interview, Jamie softens her position here: 'But halfway through the book he comes to that realisation himself, and is honest about it. So we're talking about the same thing' (qtd. in Goring n.p.).

3 As an aside, it is worth noting that Jamie has written recently about a shift occurring in nature writing in response to the burgeoning awareness of human culpability for environmental change that accompanied the publication of Rachel Carson's *Silent Spring* (1962). With this knowledge, '[w]e must [now] be meek, observant, contrite … To care about animals now we must watch them from afar' ('Diary' 39). The terms of 'connection' to the natural world, and of 'engaging' broadly understood, are surely conditioned by this commitment. My reading of *Findings* inflects a number of the ideas Jamie expresses here.

her censure is directed at an educated English male speaking for Scotland, and the tradition that she situates Macfarlane's writing in is an English one. She positions herself outside this tradition: 'I still don't like English literature as a subject' (qtd. in Goring n.p.).

If a strong narrative thread runs through the chapters of *The Wild Places*, *Findings* is more episodic. Its essays converge on common themes but might also be read alone. This too is in the service of a braiding of the wild with the domestic: encounters with the wild and the natural intersperse Jamie's everyday experience rather than standing apart from it as Macfarlane's longer ranging trips might be said to. Indeed, Jamie has suggested that favouring the shorter format of the essay has emerged out of the practicalities of combining domestic concerns with literary composition: 'over the last decade while raising my family, I've worked on shorter non-fiction' (*Kathleen Jamie* website n.p.). This is presumably also true of the poetry that Jamie has written over the same period. Again, Jamie has made the link in commenting that essays are 'affiliated to poetry' (qtd. in Goring n.p.). This affiliation (along with thematic and aesthetic alliances) informs my decision to treat Jamie's books of essays and her poetry in tandem. The broader point is that both of these formats goes some way towards presupposing a different relationship to place to that which the longer continuous travel narrative implies. If Macfarlane writes that '[i]n between my journeys, I had spent increasing amounts of time exploring the farmland and the copses within a mile or two of my home' (*Wild Places* 317), this is explicitly the case from the opening of Jamie's prose texts. The same is true of the poetry. *The Tree House* opens with an explicit braiding of the domestic and the wild. In the opening lines of the collection's first poem, 'The Wishing Tree' declares:

> I stand neither in the wilderness
> nor in fairyland
>
> but in the fold
> of a green hill
>
> the tilt from one parish
> into another.
> (3)

Emphasis is focused on a pastoral or farmed landscape ('green hills') and one that is populated and settled (with parishes). It is also a liminal space situated at the 'tilt' from one zone into another (further emphasized in this passage's enjambment). The setting is the same in *The Overhaul*. For instance, 'Glamourie' describes a landscape that 'was hardly the Wildwood – / just some auld fairmer's / shelter belt' (42). In both poems the focus is the prosaic countryside rather than the mythical wild.

From the very outset of *Findings* the domestic or quotidian perspective on the wild is established and emphasized. The text's opening view is of Black Craig, the hill visible from Jamie's home: 'The hill itself – seen from our kitchen window –

was still in silhouette, though the sky was lightening into a pale yellow-grey' (1).[4] The peak, symbol of the wild, is seen not only from within the home, but from the room most associated with domestic labour and the hearth. The tactic is repeated at the opening of the second chapter, where Jamie hears the peregrines nesting on the hill's ledges: 'The sounds enter my attic room through its window, and if I turn from my desk to glance out of that window I see the hill. She has been calling from up there' (29). As with the landform, the peregrine (a relatively rare bird of prey recently increasing in numbers after a serious decline) is a totem of the wild, framed within a domestic scene. The difference from Macfarlane's writing here is in part due to location – Jamie lives in Fife, Macfarlane in Cambridge – but it is also one of different tones of engagement. Here, the bird is heard rather than seen (a difference that Jamie foregrounds throughout *Findings*, where a range of senses have a role to play), and it is heard from the attic room where, we are told, Jamie writes. The attic room also appears as a space for thinking and reflection amidst domesticity in a number of poems, such as 'Moon' (*Overhaul* 39). The point implicit in the passage from *Findings* is that the moment of engagement is closer, temporally and spatially, to the moment of recollection and writing. Furthermore, whilst Macfarlane might write about *a* peregrine seen on a trip, Jamie writes about *the* peregrine – in this passage the female of the Black Craig pair; elsewhere she also refers to 'our local ospreys' (*Findings* 31). The birds' own domesticity is often emphasized in Jamie's terms of description. Of the peregrines she writes: 'As she called and called, he stood like a child in a huff' (*Findings* 30–31).

The same features appear Jamie's poetry. 'The Falcon' (again, note the definite article) details a young peregrine who has 'tenderly been shown / every sheltering niche' by his 'father' and 'mother'. The familial terms imply (though they do not explicitly state) a likeness between human and animal experiences. Furthermore, the scene takes place in a post-industrial landscape that bears the mark of human intervention:

> To the disused quarry
> behind our town the single
> peregrine there this May
> returns and returns.
> (*Tree House* 40)

In place of the mobile, visitor's perspective, Jamie focuses on local attachment (again, note the first-person possessive pronoun of 'our town') and signals a domesticity of perspective. As Lucy Collins notes, 'throughout her work the difference between inhabiting and visiting a place is drawn attention to by the poet' (158). In *The Overhaul*, 'Ospreys' is comparable. Here, human and avian

---

4 This pointedly domestic scene is comparable to the opening of other recent nature texts. The first chapter of Roger Deakin's *Wildwood* begins thus: 'While the rest of the world has been playing musical chairs all around me, I have stayed put in the same house for more than half my life' (3).

geographies intersect as the birds' spring migration and return to the town concludes with 'the tilt / across the A9'. Humans and animals in the poem are both 'gale battered, winter worn' in early spring. 'Ospreys' concludes by taking note of residents' attentiveness to this particular pair of birds: 'there'll be a few glad whispers round town today: / *that's them, baith o' them, they're in*' (5). This form of engagement is one of investment and care. In *Findings*, Jamie elaborates: on a stormy night, 'I had been lying awake, wondering about the birds. There are many things to fret about in the small hours, but never before have I worried about the tensile strength of an osprey's nest, atop its swaying pine' (45). Engagement is with particular birds – 'people are learning a new identification with the birds of their patch' (*Findings* 96).[5] It is on these grounds – markedly different from Macfarlane's 'epiphany of relation' – that identification occurs.

A similar pattern of identification informs Jamie's approach to ruins in the landscape. Macfarlane's reading of ruins as 'human structures from so many epochs sinking back into the land' leads to a 'sense of the human presence as being something temporary' (*Wild Places* 281). Jamie, in *Findings*, concurs that 'there's something instinctive in us, that we're drawn to human habitation and can't resist a ruin', but for different reasons: '[t]hese once-inhabited places play a different role to the uninhabited' (63). Her readings of these marks in place offer not a vision of landscape without a human figure or from which that figure has become absent, but the grounds on which to make a connection to those past inhabitants. The text repeatedly lays emphasis on the markers of domesticity in these ruins. At the remains of upland shielings, the view is '[n]ot a great vista of peaks and ridges, but a contained place, almost domestic and serene' (121). Similarly, a visit to the ruins on Ceann Iar foregrounds the domestic and labour routines that occurred there: 'There used to be plenty of people here, crofters and cottears who kept black cows, and the islands were famed for their fertility, but now there is no one left' (55). The focus on agricultural and pastoral occupations in this description emphasizes the settled nature of these places. By these means, and by the framing of her approach to the text's subject matter within her own domestic roles, Jamie establishes connections to previous inhabitants through the remnants of their presence. At Skara Brae, there is 'a huddle of roofless huts, dug half underground into midden and sand dune'. This initial description, making available a narrative of the ruins slipping back to wildness, is then subverted: here 'you can marvel at the domestic normality, that late Stone Age people had beds and cupboards and neighbours and beads. You can feel both their presence, their day-to-day lives, and their utter absence' (11). 'Marvelling' in this passage is akin to wonder, but emanates from a different mode of connecting.

If Jamie sees Macfarlane's work as inhabiting a tradition of men *alone* in the wild – which is a generalization to which there are significant exceptions, and that does not necessarily take account of his evident awareness that 'wild' landscapes

---

5 Again, Deakin provides a fruitful comparison in his relationship to *individual* trees: 'Each tree we encounter has its own story to tell' (38).

are often emptied rather than simply empty – her own texts are at pains to repopulate our sense of these places. On several occasions, *Findings* goes as far as to parody attempts to experience wild places alone. At the Neolithic tomb at Maes Howe, where the undermined 'gesture of connection' discussed above occurs, Jamie emphasizes the thwarting or disappointing of such modes of engagement. Expecting a sombre, dark atmosphere inside the tomb, Jamie is surprised to find a pair of surveyors with their equipment and bright lighting. This circumstance prompts reconsideration:

> Something interesting was happening here. It occurred to me … that I would never again get so close to real Neolithic ancestors. Had this scene not happened before, thousands of years ago? Had not skilled workmen stood within this very tomb at the end of the working day, and taken a moment to survey their handiwork? (19–20)

In the similarity between past and present labour in this place, what was the distraction has become the means of connection. A passage describing an archaeological dig in *Sightlines* makes an analogous comparison: 'The director in his report calculated that a hundred people would have been required to build the henge, probably organised in gangs. A similar team, organised in gangs over two seasons, dismantled it' (56). Hence, 'to abolish all layers of time took only a little imagination' (58). Again, *The Tree House* contains similar examples. In the suggestively titled 'The Reliquary', Jamie writes that '[t]he land we inhabit opens to reveal / the stain of ancient settlements, / plague pits where we'd lay / our fibre optic cables' (37). The imposition of a modern communications infrastructure is the means by which the marks of ancient lives are discovered (rather than destroyed). It is noteworthy that 'inhabiting' (rather than 'visiting') a place makes such observations available.

Throughout *Findings*, it is human-avian interactions that offer the most sustained braiding of the wild and the domestic. Orkney's 'fulmar tenanted cliffs' are not separated from but echo the marks of pastoral human inhabitation also found there: 'The islands have been farmed for a very long time, and if you climb a hill and look down onto the green lands below, the farmhouses are so plentiful they look like as though they've been shaken out from a box' (7–8). Cohabitations are, again, noted from a distinctly domestic perspective:

> I was fetching the children's school clothes out of the tumble dryer, when I chanced to look out of the window and saw the male osprey. … He looked as though he was trying to get enough lift to carry him over the summit. There's a fishing loch over there stocked with trout, a fast-food takeaway for a bird like this. Then the osprey was gone and I turned back to the dryer, looking for matching socks. (46)

The final two sentences quoted here balance natural and domestic foci. The wild hunting bird fishes in a loch 'stocked' by and for humans ('Ospreys', in *The Overhaul*, also incorporates this observation). The tasks of everyday life – washing, feeding – are shared between species. Jamie goes on to note that 'I like being able to look up from my everyday business, to see the osprey or the peregrine going about hers' (*Findings* 46).

The ideal to which Jamie aspires, then, is a *partial* perspective in two senses: one that is interested (or perhaps invested) in particular locales and animals, and one that is turned towards the wild and the natural but acknowledges its own domesticity. 'The Tree House' – as the title poem of the collection a major statement – describes this abode as somewhere on the edge of domesticity (in the garden but certainly separate from the house), situated 'just out of reach' with a view 'over house roofs', where the narrator temporarily escapes 'our difficult / chthonic anchorage' to make 'a dwelling of sorts' (41–3). Again, in *Findings*, an avian connection describes this marginal position: 'Chimneys are the jackdaws haunt, and I envy them their elevated life, at once part of a household and part of the wide air' (99). Jamie's own mode of interaction with nature emulates this marginality: 'I keep my binoculars about me, and catch a glimpse at coffee-time, or before fetching the kids from school' (38). Indeed, marginality receives extended treatment in a passage that draws together the two domestic activities already detailed:

> Between laundry and fetching the kids from school, that's how birds enter my life. I listen. During a lull in the traffic, oyster catchers. In the school playground, sparrows – what few sparrows are left – chirp from the eaves. There are old swallows' nests up there. It's late April, but where are the swallows? The birds live at the edge of my life. That's okay. I like the sense that the margins of my life are semi-permeable. Where the peregrines go when they're not at their rock ledge, I couldn't say. (39)

Permeability carries at least two resonances here. The passage expresses the shared environment of humans and animals: the sparrows can be heard in the eaves of the school, the swallows' nests are also built on a man-made structure. Other passages of *Findings* demonstrate the same point: at Ceann Ear's abandoned villages, 'Fulmars nest in the corners now, and there are nests of cheeping starlings in the crannies of the stonework' (65). Furthermore, Jamie's mode of engagement is one that breaks down (permeates) boundaries between experiencing nature and everyday life – it exists at the edges of routine, acting almost as its enlivening counterpart.

Listening, emphasized here, also has a role to play. This passage follows a discussion of 'a tradition of lone men engaging with birds' that focuses on watching (38). Jamie's aural engagement foregrounds a distinction to be made between seeking out nature and allowing it, as here, to 'enter one's life'. Notwithstanding the fact that '[a]s humans, we privilege sight' (94), Jamie is repeatedly invested in listening: she writes about appreciating 'the company of people who listen to the world' (54). By these means, a shift in the balance of agency effaces the imputation of (visually constructed) meaning in Jamie's writing. The aural plays a similar role in *Findings* as the haptic does in *The Wild Places* (which, incidentally, is less interested in sound) – both texts aim to re-evaluate the dominance of scopic experience in inherited traditions. Jamie's work, like Macfarlane's, is thus clearly engaged with phenomenological positions on landscape and a desire to look beyond Cosgrove's 'ways of seeing' (55). The places articulated in this manner are certainly practised rather than only symbolic.

## Aesthetics of Reticence

Jamie's prose displays a marked reticence around speculating about nature and wild animals beyond what can be observed. Examples abound in the last few quoted passages. In the 'how birds enter my life' passage, Jamie 'couldn't say' where the peregrines go when they're out of sight. The question of the swallows' whereabouts considered just before this assertion also goes unanswered. Indeed, the passage as a whole is characterized by doubling back on previous observations in short, clipped phrases that, collectively, create an aesthetic of reticence or humility in the description of nature. It can also be seen in the final sentence of the passage describing the prospect from the laundry: 'Then the osprey was gone and I turned back to the dryer, looking for matching socks' (*Findings* 46). The scene of human/avian connection is closed abruptly with the return to domestic routine as soon as the bird is lost to sight. The same emphasis on resisting imposition of meaning is discernible throughout the text. Jamie is explicit in acknowledging that she 'sail[s] on the surface of understanding, a flicker here, a silence there' (*Sightlines* 17); and she finds a tone to match. Describing the process of composing these essays, Jamie emphasizes the attention she pays to aesthetics of syntax: 'I will play with the grammar, do quick handbrake turns with the subject, or put in a two-word sentence to pull it up sharp.' In this respect, the process is akin to her poetry writing: 'I will change a sentence to alter the assonance and consonance, the same way I would with a line of poetry' (qtd. in Goring n.p.). It follows, then, that Jamie's poetics are equally spare, and for the same reasons. This is also amply evident from the poetic excerpts quoted already. This aesthetic evidently emerges from Jamie's commitment with respect to her entire published output that 'it's all provisional' (*Kathleen Jamie* website n.p.). Thus, the subtitle appended to the North American edition of *Sightlines* – 'A Conversation with the Natural World' – is apt in signalling an ongoing dialogue left open. Jamie's texts – in poetry and prose – are attentive to nature, but also aware of the limitations against which that attentiveness presses. They are written in a style careful not to cross that line.

As a result of this awareness, Jamie's texts invoke and manipulate an ever-varied set of cadences attached to words such as 'viewing', 'observing', and 'watching'. They also encompass passages that describe and analyze the relationship between observing and recollecting. In some instances, as in the opening passage of *Findings* discussed above, there is an attempt to diminish the distance between the two processes. Elsewhere, there is a conscious effort to hold them apart. On a walk to see the local ospreys' nest, Jamie notices an unusual bird overhead and (in the retrospective account given in the text) tells herself 'to be calm, to look carefully and remember' (41). Similar self-instruction appears in *Sightlines* when Jamie describes how the 'mineral silence' of icebergs is 'deep and frightening, and makes my mind seem clamorous as a goose'. She consciously attempts to 'quell my mind' in response (4). In the passage from *Findings*, the initial description of the bird, later identified as a crane on an uncommon migration path or simply lost, is anatomized and undertaken by analogy: 'it was like a brittle flying stick, side-on, growing thicker as it neared'; '[i]n profile, I could see clearly the head held far out on a neck so long and fragile it might snap, and the long legs and feet extended just as far behind' (41). The effect – comparable to defamiliarizing techniques in Macfarlane's writing – is to resist the classificatory impetus of natural history writing. Instead, the moment of engagement, of seeing, is extended in a passage replete with present participles:

> This is what I want to learn: to notice, but not to analyse. To still the part of the brain that's yammering, 'My god, what's that? A stork, a crane, an ibis? – don't be silly, it's just a weird heron.' Sometimes we have to hush that frantic inner voice that says 'Don't be stupid,' and learn to look, to listen. You can do the organising and redrafting, the diagnosing and identifying later, but right now, just be open to it, see how it's tilting nervously into the wind, try to see the colour, the unchancy shape – hold it in your head, bring it home intact. (42)

'Hushing the inner voice' here is comparable to other attempts to efface projections onto place discussed elsewhere in this monograph. Again, Jamie is seemingly resisting scientific classificatory models of engagement in this passage ('diagnosing' and 'identifying'), but there is also a recalibration of more poetic types of cerebral processing ('redrafting' is certainly more writerly in scope). Wordsworthian recollection-in-tranquillity is refracted here with an altered weighting of emphasis: noticing in the moment is held apart from (and, to some extent, privileged over) thinking through in retrospect. That these concerns are central to the whole text is indicated by its title: the ambiguity of 'findings' (a plural noun from a present participle of a verb), in its reconsideration of scientific discourses of knowledge about nature, expresses the difference between and relation of the moments of experiencing and representing (to one's self in remembering, and to others in subsequent writing). If this passage signifies a refusal to impose meaning in the moment, it is moderated in resisting a '*frantic* inner voice' – it is the agitated tone of thought that is to be exorcised rather than thought *per se*. Elsewhere, the description is also coloured rather than simply neutral: 'unchancy' might be regarded as only a figurative description of the crane's flight, but 'nervous' seems more anthropomorphic. The relationship between observation and recollection is reconfigured rather than refuted in Jamie's text: attentiveness might be awarded more prominence than heretofore, but 'redrafting' is not simply abandoned.

Whilst this emphasis on noticing and observing might appear to be a point on which Jamie's writing can be distinguished from that of Macfarlane, wonderment – associated to this point with the latter and not the former – also has a role to play here. Jamie's terms of description often imbue attentiveness with a kind of wonder. When Jamie first sees the crane in the above passage, she describes having 'knelt down on the earth to make myself small' (41). In practical terms, such an action enhances the chances of the bird flying overhead undisturbed and allowing a better view; figuratively, it signifies the removal from view (or 'making small') of self that she is interested in here as elsewhere. However, there is also another register to this that recurs throughout the passage: 'Like some medieval peasant granted a vision, I was kneeling in a field, fixated by this uncanny cross in the sky' (41). This attitude of supplication inscribes a religious aspect to attentiveness. The 'vision' seen is complex: it refers to the sensory engagement with nature that the passage goes on to emphasize, but this literal meaning is also supplemented by a figurative one and employs a stock literary archetype to suggest a reverential aspect of the relationship to nature. These terms of description suggest a moment of epiphany, an 'epiphany of relation' even, that, though accented differently, is comparable to those found in Macfarlane's text. The language of reverence also infuses the domestic scene of many poems where, for example, in 'Doing Away', Jamie writes that there is 'nowhere I'd rather be / than here, fulfilling *my daily rituals*' (*Overhaul* 9, my italics). As *Findings* progresses, increasingly explicit links are made between attention and prayer: noticing 'with intensity' is the optimum to which Jamie aspires (113), her text culminating in a whale-watching trip that 'proceeds like a kind of theology – by glimpses, sightings' (186). Asked by a friend if she prays, Jamie says she does not, but that she 'noticed, more than noticed'. She continues: 'Isn't that a kind of prayer? The care and maintenance of the web of our noticing, the paying heed?' (109). In *Findings*, the answer is undoubtedly yes. This is another version of wonderment at nature and place.

'More than noticing' certainly opens up a space in which representation operates in tandem with attentiveness to the moment of engagement. If the latter is emphasized in Jamie's work, the role to be played by representation of place in writing is implicitly problematized – it is associated with the mediation and imposition that she is concerned to set limits on. Indeed, these limitations are an explicit presence in her texts. A contrast is drawn between Jamie's role as a creative writing lecturer – 'my university students were … expecting to be taught to engage with the world in language' (*Findings* 165) – and the attraction of places precisely because they evade literary or even linguistic expression. The act and aesthetics of describing are at stake here and throughout. This is the case on the cliffs of Lewis:

> I sat on a damp rock, took my notebook from my inner pocket, and made earnest notes … I made notes, but the reason I'd come to the end of the road to walk along the cliffs is because language fails me there. If we work always in words, sometimes we need to recuperate in a place where language doesn't join up, where we're thrown back on a few elementary nouns. Sea. Bird. Sky. (164)

In this passage, 'recuperation' is treated differently than in other instances noted thus far. Language – the habitual medium of representation – is incapable here but for 'a few elementary nouns'. Recuperation occurs outside of linguistic parameters. The related problem of reading 'markings' in the landscape appears elsewhere in the text. Of Neolithic stone carvings, Jamie writes:

> I didn't know what to make of them, and in truth felt a bit daft, hunkering in a field like this, so I made the obligatory musings about the Neolithic hand which made them thousands of years ago, wondered what they had been for, inscribed for. What they meant, if indeed 'meant' was quite the right word. The carvings themselves offered no clue, but lay mute in the morning sunshine. (116)

For Jamie, the 'mute' carvings refuse the 'obligatory musings' that define a habitual desire to find meaning. Here, the idea of meaning in this sense is called into question – the type of 'wonder' with which it is associated is undermined. Instead, the limits of representation are respected and allowed to register as such in the texts.

Several poems also dramatize the incapability of language and the superfluity of the attempt in the face of nature's own articulations. 'The Dipper' describes an encounter with this species of bird at a waterfall in which the narrator listens to the flow of its 'supple, undammable song'. The final stanza then question's poetry's ability to replicate the bird's voice:

> It isn't mine to give.
> I can't coax this bird to my hand
> that knows the depth of the river
> yet sings it on land.
> (*Tree House* 49)

This is the final poem of *The Tree House* and these lines therefore signify its closing statement. The acknowledgement of the limits of language is thus all the more emphatic in marking a silence that goes unbroken in this collection. The same theme appears again in 'May' from *The Overhaul*. Here, a blackbird's song is 'telling us what he thinks of it [the month of the title and the new season]'. The poem is a sonnet with a volta turning attention in its second half to the same problem of poetry:

> What can we say
> the blackbird's failed
> to iterate already? Night calls:
> the windows of next-door's glass house
> crimson, then go mute.
> (7)

The question – how might poetry replicate what the bird already says exactly – receives no answer. The poem resorts to the indirectness of symbolically loaded description and then ends in going 'mute'. The experience itself, and the difficulty of recording that experience, are therefore what the poem registers. Elsewhere, even where this dilemma is not treated, the same awareness can be felt in the pared-back description and absence of explaining commentary. 'Landfall' describes a scene comparable to that on the cliffs quoted from *Findings* in the previous paragraph. Its six lines comprise five that record what is noticed 'when we walk at the coast', in this case a single migrating swallow making landfall, followed by one that asks, in light of this observation, 'can we allow ourselves to fail' (*Tree House* 15). The haiku-like poem draws a lesson from simple subject matter but pointedly does not say more. Its reticence is respectful, in line with the questions posed in 'The Dipper' and 'May'. Jamie has suggested that 'to write a poem is to work with language until, as near as we can get it, the true shape, the truest utterance of that poem, is arrived at' ('Truth and Change in Poetry' 280). More often than not in her verse, working with language is removing superfluity. The truth that emerges takes the form of elemental description.

Despite these reservations, Jamie is, by the evidence of the text's existence, invested in the value of writing about place and nature. Despite previously thinking that 'language was what got in the way', and that 'you could not get at the world because you were stuck with language', Jamie has spoken about revising these ideas: 'Now I think language is what connects us with the world' (qtd. in Scott n.p.). Louisa Gairn has noted this aspect of Jamie's writing: she suggests that Jamie is one of a number of contemporary Scottish writers who 'are not only reviewing human relationships with nature, but also the role writing has to play in exploring and strengthening that relationship – helping to determine the ecological "value" of poetry and fiction' (156). Indeed, *Findings* includes formulations attending to the role of literature in this respect: 'It's poetry's job, isn't it, to keep making sense of the world in language, to keep the negotiation going? We can't relinquish that' (177). For Jamie, writing about place must overcome this bind by demonstrating and remaining aware of its own processes just as, my analysis shows, occurs throughout her poetry and prose. This alertness is detailed in a passage from *Findings* concerned with the techniques and uses of representation. In particular, like Sebald and Carson, Jamie compares her practices as a writer to those involved in photography. Watching salmon leaping rapids on their way back to spawning grounds on the river Braan, Jamie happens upon a photographer instructing students on a course. His instruction, like her creative writing teaching, would focus on technique 'to make a better photograph of the natural world'. He would demonstrate '[h]ow to master all that technical stuff but make the image look fresh, natural, accidental even. How to employ all that technique expressly to make the result look natural, techniqueless' (75). Technique, at least initially, appears to be that which distances forms of representation from the natural subject matter. However, Jamie subsequently notes how she 'began watching differently the expressionless fish. I was watching for technique' (75). From this altered perspective, she sees that the salmon themselves employ technique in their leaping, even if its workings are 'hidden below the water' (76). Indeed, that the technique remains hidden aligns it with that of the photographer and, by implication, that of the writer – of Jamie herself. Technique, then, and particularly in the present context, the techniques of language, offers grounds for a connection between nature and its representation as opposed to the feature that distinguishes them. Thus, Jamie asserts elsewhere that 'language is what we do as human beings, that's where we're at home, that's our means of negotiating with the world. So it doesn't get in the way it enables. We do language like spiders do webs' (qtd. in

Goring n.p.). Or, to insert the former analogy, like salmon do leaping. Language, though a cultural form, is, in the final analysis, what we do naturally.

Jamie, then, is concerned to write texts that are alert to the potentials and limitations set by literary form and their status as texts bearing the marks of artifice and mediation. She is attuned to those concerns for the aspects of experiencing place and nature that escape or exceed textual representation – the same concerns that this monograph attempts to reconcile – but also dramatizes arriving at understandings of the contribution representational technique and construction might play in a nuanced sense of place. Responding to the dilemma of imposing place's meaning, Jamie's writing is characterized by a humility that locates it closer in style to reportage than testimony. In a telling construction, she has said that in *Findings* her aim was to 'get the ego out of the way and just look and see what's there' (qtd. in Scott n.p.). Macfarlane suggests that '[t]he finest writing about landscape is almost always modest, exact and attentive. It is precise without ever being curt' ('Common Ground' n.p.). He might easily be describing Jamie's writing in identifying these qualities. In *Findings*, literary technique occurs 'hidden below the water' (to borrow a metaphor from the passage above). If Jamie is concerned to reconsider ideas of intimacy with place and nature, she is also concerned to recalibrate the terms in which that intimacy is described.

Jamie's texts tell narratives that shift the definition of 'wild' away from wilderness untouched by human intervention and towards a concept of place that recognizes a braiding or entwining of human and non-human forces in the determination of landscapes. If this shift is towards the human in terms of the production of space, it is measured against a coterminous concern to remain aware of the investments and dangers attending to an anthropomorphic centralization of the human in speaking for or representing place. *Findings* and *Sightlines* overtly move away from this tradition of nature writing. These texts demonstrate an awareness of and responsiveness to the set of concerns that are presently engaging academic considerations of place with which this book opened and to which it now returns.

# Afterword
# Testifying to Place

Discussing a pictorial tradition of landscape, Kenneth Clark asserts that '[r]ealistic landscape, which the ignorant believe to be one of the easiest forms of painting, is actually one of the most inaccessible, one in which success is rarest and most precarious' (101–2). A merely transcriptive aesthetic, I wish to suggest, is equally problematic in *literary* landscapes. The extent to which an aesthetic privileging the experiential approximates transcription (and, indeed, whether or not this is its aim) comes under scrutiny here. The final two case studies of this monograph, those approaching the nature writing of Robert Macfarlane and of Kathleen Jamie, emphasized most clearly some of these issues that are more or less implicitly present throughout the preceding chapters. Of the section headings under which chapters have appeared, 'engaging' is perhaps the most open. To engage (as with to picture, to map, to walk, to notice) necessarily implies the activity of the subject in an articulation of place, but it is not intended that this highlighting is necessarily to downplay other more passive forms of presence that might seem to more readily approximate transcription. One can engage 'in' place as well as 'with' place. The final cases studies, with their similar set of concerns, thus flow into this epilogue. The interplay between the subjectivity of accounts of landscape that foreground an experience of place, and the transcription, description, and/or depiction of that place, is complex. Whilst it is with this issue that my monograph closes, it is not my aim to offer closure, but rather to point towards some of the directions currently being taken in the ongoing development of contemporary literary landscapes.

Lorimer and Wylie assert that '[t]he landscape is a multiplicity of narratives and perceptions, and you can lose and find yourself among this multitude' (7). In its selection of texts from a wide and diverse field of contemporary landscape writing, this monograph has conferred attention on writers who emphasize rather than elide the subjectivity of their responses to place. In the course of reviewing *The Ground Aslant: An Anthology of Radical Landscape Poetry*, edited by Harriet Tarlo and published in 2011, Robert Macfarlane testifies to the fact that marking subjectivity is not a universally applied technique for the literary representation of place:

> The lyric 'I', the first-person witness and narrator so central to prose writing about nature, barely survives here. … There is, for sure, a long tradition of writing about loss-of-self in nature: Richard Jefferies' sunlit deliquescence on a Wiltshire hillside, for instance, or Nan Shepherd's 1945 account of 'melting' into Cairngorm granite. Here, though, nature offers neither a compensatory completion in the Mother-other nor a Zenishly austere dispersal. Rather, the 'I' gets aggressively dismantled. … Landscape 'detail' provides no reliable resting place for the eye or the mind. It simply refers the subject onwards in an effortful relay of attention from speck to speck. ('Meshed and weathered' 14)

Despite the absence of the lyric 'I', Harriet Tarlo, in her introduction to *The Ground Aslant*, testifies to the fact that '[i]ntimate observation of and involvement with a particular place remains at the heart of much landscape poetry, radical or otherwise' (13). The difference, if it is to be upheld, is one of descriptive aesthetic rather than mode of engagement. If Macfarlane's typology of lyric prose and impersonal poetry is overly schematic (and my study in this monograph of Carson's and Jamie's poetry suggests that it is), these remarks are nonetheless informative in a number of interesting ways. If a literary transcription of landscape (allied to that which Clark finds to be so elusive in pictorial representation) consists in a 'relay of attention' from feature to feature, then the marking of subjectivity in the lyric 'I' would seem to have different ends in view. Indeed, just as Macfarlane suggests an inherited tradition of the 'loss-of-self' aesthetic running through Jefferies and Shepherd to contemporary avant-garde poetry, the implicit reference to his own 'lyric' antecedents suggests another lineage playing into the contemporary recounting of subjective experience of place. As Tarlo's introduction makes clear, the poets collected together in *The Ground Aslant* aim to displace what she calls 'the great romantic myth of originality, of the poet as a genius' (15). For Macfarlane, and for others studied here, the break is not as clear cut: the reference to lyricism is inevitably caught up in a set of inherited concerns. Writing not of place *per se* but of an experience of place in a sensuously responsive manner is, as Macfarlane hints, a distinctly lyric mode. The contemporary inheritance from lyric traditions, the Romantic foremost amongst them, is one that, in various ways, reconfigures the balance of agency between place and writer.

The lyric subjectivity that features in what I have called landscape writing here is perhaps a particular manifestation of a more widespread and important trend in contemporary literature. David Shields's recent polemic, *Reality Hunger: A Manifesto* (2010), has championed the 'lyric essay' as a 'hybrid, shape-shifting, ambiguous' form that 'makes a flow chart of our experiences of the world' (31). Sebald features amongst Shields's examples of writers who have written texts 'in which "nonfiction" material is ordered, shaped, and imagined as "fiction"' (14). If Sebald and others studied here emphasize experience of place, that experience is focused in and channelled through a particular narrating persona. Whilst the privileged position of narration that this lyricism endows might seem to allow the writer to speak for place in an authoritative way, I want to suggest that (notwithstanding the implications of publishing texts) this is not the case in the contemporary writing of place and experience addressed here. Rather, the writers with which this monograph is concerned mark the subjective rendering of place that their texts undertake, at least in part, to deflate or undermine any potential claims to an authoritative or definitive vision. Thus, for example, in *London Orbital*, Sinclair asks, 'Why let someone else nominate where is worth visiting? … Don't take my word for it, don't bother with my list of alternative attractions … discover your own. In the finding is the experience' (*London Orbital* 318–19). In the foregrounding of experience of place, a more democratic process of formulating meaning emerges. These writers, it can be inferred, do not consider textual representations to fix or freeze meaning. Rather, they employ an experiential aesthetic to recuperate moments in the fluid (open and ongoing) production of place.

This reading is supported by Macfarlane's conflation of the 'lyric "I"' and 'the first-person witness and narrator'. These texts bear witness to, or offer testimonies of, experiences of place. 'In its most traditional, routine use in the legal context', Shoshana Felman and Dori Laub note, 'testimony is provided, and is called for, when the facts upon which justice must pronounce its verdict are not clear.' Testimony in general thus does *not* offer 'a completed statement, a totalizable account'; but rather, in it, 'language is in process and in trial, it does not possess itself as a conclusion, as the constatation of a verdict or the self-transparency of knowledge. Testimony is, in other words, a discursive *practice*, as opposed to a pure *theory*' (5–6). Similarly, for Derrida, testimony is defined by its subjectivity: 'if the testimonial is by law irreducible to the fictional, there is no testimony that does not structurally imply in itself the possibility of fiction, simulacra, dissimulation, lie, and perjury'. Thus:

> If this possibility that it seems to prohibit were effectively excluded, if testimony thereby became proof, information, certainty, or archive, it would lose its function as testimony. In order to remain testimony, it must therefore allow itself to be haunted. It must allow itself to be parasitized by precisely what it excludes from its inner depths, the *possibility*, at least, of literature. (*Demeure* 29–30)

The shared aesthetic emphasizing experience in the texts of Sebald, Carson, Sinclair, Macfarlane, and Jamie, expresses this possibility *not* as a danger, but rather as a positive feature. This, I suggest, is the function of a combinatory approach that looks to situate testimonies of individual experiences (that is, individual in the sense of both singular and personal), within the context of a history of representations of a place.

This approach to writing place has not been universally taken up in contemporary writing. Alongside poetic departures from the lyric, a number of prose writers have sought to efface subjective projections. Paul Farley and Michael Symmons Roberts's *Edgelands: Journeys into England's True Wilderness* (2011) is paradigmatic of one approach to the considerations of contemporary landscape writing in its implicit critique of Macfarlane and others like him. The edgelands that Farley and Symmons Roberts describe are 'a difficult landscape to immerse ourselves in physically – there would be no tree climbing, and swimming in standing water was out of the question – though in the back of our minds there was a sense of letting the terrain speak for itself, rather than framing ourselves within it as intrepid explorers' (9). A similar apparent objectivity is also described in a recent text by Olivia Laing. *To the River: A Journey Beneath the Surface* (2011) models a process of 'ecstatic surrender', wherein a 'sort of dislocation with the human, identifying instead with dead or lifeless matter, seems both a natural and necessary part of becoming absorbed in the wider realm' (199). The extent to which either of these texts, or any text, lets the terrain speak for itself is an unresolved question both for creative writers and in debates across academic disciplines. Despite these stated objectives, over their course both books seem heavily freighted and inflected with subjective positions and impositions. For Laing, in this passage, the reverie-like tone of '*ecstatic* surrender' offers a mitigating device. My aim here is not to draw a contrast between two different kinds of project, but to suggest that the writers I have dwelt on at length have accented a critical self-awareness of the investments made in landscape writing. They have not sought to efface subjectivity, but to qualify it. The texts of Sebald, Carson, Sinclair, Macfarlane, and Jamie bespeak an understanding that letting the terrain speak for itself might remain as inaccessible in textual form as Clark finds it to be in a pictorial tradition. An experiential articulation of landscape and place circumvents the need to make any such claim: it charts a more circumspect and, in some respects, a more nuanced course through these problems. The immersion that occurs there is physical *and* textual, embodied *and* representational, and, above all, partial.

The related issues of verisimilitude and veracity that are raised by Sebald's 'picturing' of landscape, and in particular by his engagement with the purported objectivity of photographic recordings of place, play out across a number of references to that medium of representation made by writers featuring in this monograph. Jamie's discussion of the 'technique' of the photographer taking pictures of leaping salmon, with which Chapter 5 concluded, addresses similar concerns. These references, then, along with a few others, offer the grounds for a gathering together of findings here. Carson describes his aim in writing about Belfast, in ambiguous terms, as being 'to act as a camera or a tape-recorder, and present things in a kind of edited surreality' (qtd. in Brown 148–9). For Carson, even in attempting to replicate the documentation of technological apparatus, the shift from reality to 'surreality' that results from editing – involuntary as well as intentional – is inevitable. The result, Carson suggests elsewhere, is that no one account of place is definitive: 'Anything can be told this way or that. There's no final way of telling a story, or explaining the totality of whatever it was that happened at any given time' (qtd. in Brandes 84). Closure and exhaustive coverage are both squarely undermined in this statement. Instead, narratives contribute to ongoing processes of determination from particular perspectives.

Iain Sinclair's oxymoronic description of his text *Hackney, That Rose-Red Empire* as 'a documentary fiction' plays off the same instabilities of recording (579). In a passage from *Ghost Milk: Calling Time on the Grand Project* (2011), he describes his 'snaps', photographs taken on walks, as 'part of a logging process, the laying down of an archive from which a more mediated account would be teased' (80). Sinclair, like Carson, is perhaps playing off Christopher Isherwood's famous declaration – following on immediately from the opening description of street scene in *Goodbye to Berlin* (1939) – that, as a writer, he is 'a camera with its shutter open, quite passive, recording, not thinking' (9). Both Sinclair and Carson eschew that passivity and foreground subsequent layers of mediation instead. Furthermore, Sinclair goes on to suggest that photography does not always constitute this simple act of logging. Robin Maddock, a Hackney-based photographer, is 'a troubled witness coping with the responsibility of shaping a true report. The person you never see, the one with the camera, is omnipresent; in the way that, however alien the set, every portrait becomes a self-portrait' (*Ghost Milk* 80–81). The testimony of a 'witness' is necessarily subjective. Sinclair's 'mediated account' is Carson's 'edited surreality' – both complicate simple notions of reportage. Like Carson, Sinclair works from this position to draw attention to the partiality of the texts that emerge. 'The fiction of place', he writes, 'is dictated by many voices, physical structures without number, the visible and the invisible.' In this echo chamber, Sinclair's text is 'a story of fallible memory, inaccurate or inventive transcriptions, hard-earned prejudices, false starts and accidental epiphanies' (*Hackney* 579). If there is a certain amount of (ironic) mysticism in this channelling of the energies of place, it only serves to reinforce the point: these accounts are not to be thought of as definitive versions of places but as subjective testimonies intervening in their making. Perspective inheres in description.

The recourse to the terminology of fiction and story in these statements facilitates a final comparison. Cultural geographer Doreen Massey has described space as a 'simultaneity of stories-so-far' (*For Space* 9). For Massey, 'an insistence on the contemporaneous multiplicity of stories' challenges any notion of 'landscape as a oneness': 'To walk across a landscape with any degree of awareness is to pick your way across the locations of a host of unfinished trajectories. Their unfinishedness addresses our today' ('Landscape/space/politics' n.p.). Places, thought of in this way, Massey writes elsewhere, 'necessitate intervention; they pose a challenge. … They require that, in one way or another, we confront the challenge of the negotiation of multiplicity' (*For Space* 141). The modes of landscape writing taken up by a number of contemporary writers go some way towards answering this challenge. They share a combinatory approach to writing landscapes that draws on experiential, practised engagements with place, but also situates this first-hand knowledge within longer literary and pictorial traditions and histories of representing those same places. Sinclair writes: 'I've never been much good at recognizing a division between fiction and reality, past and future, this place and that place' (*Ghost Milk* 138). The texts studied here do not attempt to pull apart these components, but rather to recognize the entwining of this experienced place with that represented one. As such, each of these texts articulates a story that enters into the accretive multiplicity of those landscapes so far.

# Works Cited


Alcobia-Murphy, Shane. *Governing the Tongue in Northern Ireland: The Place of Art/The Art of Place*. Newcastle upon Tyne: Cambridge Scholars Publishing, 2005.

Alexander, Neal. *Ciaran Carson: Space, Place, Writing*. Liverpool: Liverpool University Press, 2010.

———. 'Mapping Junkspace: Ciaran Carson's urban cartographies'. *Textual Practice* 21.3 (2007): 505–32.

Alexander, Neal, and David Cooper, eds. *Poetry and Geography: Space and Place in Post-war Poetry*. Liverpool: Liverpool University Press, 2013.

Ammons, A.R. 'A Poem is a Walk'. *Epoch* (Fall 1968): 114–19.

Anderson, Ben, and Paul Harrison, eds. *Taking-Place: Non-Representational Theories and Geography*. Farnham: Ashgate, 2010.

Augé, Marc. *Non-Places: Introduction to an Anthropology of Supermodernity*. Trans. John Howe. London: Verso, 1995.

Baker, Brian. *Iain Sinclair*. Manchester: Manchester University Press, 2007.

———. 'Maps of the London Underground: Iain Sinclair and Michael Moorcock's Psychogeography of the city'. *Literary London: Interdisciplinary Studies in the Representation of London* 1.1 (2003): n.p. Web. 14 July 2010.

Baker, J.A. *The Peregrine*. New York: NYRB Classics, 2005.

Banville, John. 'The Rubble Artist'. *The New Republic* 225.22 (2001): 35–8.

Barry, Peter. *Contemporary British poetry and the city*. Manchester: Manchester University Press, 2000.

———. '"You can get it from the street": the prose style of Iain Sinclair'. *City Visions: The Work of Iain Sinclair*. Ed. Robert Bond and Jenny Bavidge. Newcastle: Cambridge Scholars Publishing, 2007. 44–55.

Barthes, Roland. *S/Z*. Trans. Richard Miller. Oxford: Blackwell, 1990.

Bate, Jonathan. *The Song of the Earth*. London: Picador, 2000.

———. 'To the River by Olivia Laing – review'. *Guardian* 18 June 2011: n.p. Web. 18 July 2011.

Benjamin, Walter. *Charles Baudelaire: A Lyric Poet in the Era of High Capitalism*. Trans. Harry Zohn. London: Verso, 1973.

———. *Illuminations*. Ed. Hannah Arendt. London: Pimlico, 1999.

Bennett, Jane. *The Enchantment of Modern Life*. Princeton: Princeton University Press, 2001.

Berberich, Christine. 'England? Whose England? (Re)constructing English Identities in Julian Barnes and W.G. Sebald'. *National Identities* 10.2 (2008): 167–84.

Berger, John. *The Look of Things*. New York: Viking Press, 1974.

Blair, Sara. 'Cultural Geography and the Place of the Literary'. *American Literary History* 10.3 (1998): 544–67.

Bond, Greg. 'On the Misery of Nature and the Nature of Misery: W.G. Sebald's Landscapes'. *W.G. Sebald: a critical companion*. Ed. J.J. Long and Anne Whitehead. Edinburgh: Edinburgh University Press, 2004. 31–44.

Bond, Robert. *Iain Sinclair*. Cambridge: Salt Publishing, 2005.

Bonnett, Alistair. 'The Dilemma of Radical Nostalgia in British Psychogeography'. *Theory, Culture & Society* 26 (2009): 45–70.

Borges, Jorge Luis. *Labyrinths: Selected Stories and Other Writings*. Ed. Donald A. Yates and James E. Irby. London: Penguin, 1970.

Brace, Catherine, and Adeline Johns-Putra, eds. *Process: Landscape and Text*. Amsterdam, Netherlands: Rodopi, 2010.

Brandes, Rand. 'Ciaran Carson interviewed by Rand Brandes'. *The Irish Review* 8 (1990): 77–90.

Brearton, Fran. 'Poetry of the 1960s: the "Northern Ireland Renaissance"'. *The Cambridge Companion to Contemporary Irish Poetry*. Ed. Matthew Campbell. Cambridge: Cambridge University Press, 2003. 94–112.

Bridgeman, Teresa. 'Time and Space'. *The Cambridge Companion to Narrative*. Ed. David Herman. Cambridge: Cambridge University Press, 2007. 52–65.

Brooker, Peter. 'Iain Sinclair; The psychotic geographer treads the border-lines'. *British Fiction of the 1990s*. Ed. Nick Bentley. Abingdon: Routledge, 2005. 226–37.

———. *Modernity and Metropolis: Writing, Film, and Urban Formations*. New York: Palgrave Macmillan, 2002.

Brooker, Peter, and Andrew Thacker, eds. *Geographies of Modernism: Literatures, cultures, spaces*. London: Routledge, 2005.

Brooks, Peter. 'Freud's Masterplot'. *Literature and Psychoanalysis: The Question of Reading, Otherwise*. Ed. Shoshana Felman. Baltimore, MD: Johns Hopkins University Press, 1982. 280–300.

Broom, Sarah. *Contemporary British and Irish Poetry: An Introduction*. Basingstoke: Palgrave Macmillan, 2006.

Brosseau, Marc. 'The city in textual form: *Manhattan Transfer*'s New York'. *Ecumene* 2.1 (1995): 89–114.

———. 'Geography's literature'. *Progress in Human Geography* 18.3 (1994): 333–53.

Brown, John. *In the Chair: Interviews with Poets from the North of Ireland*. Cliffs of Moher: Salmon Publishing, 2002.

Bulson, Eric. *Novels, Maps, Modernity: The Spatial Imagination 1850–2000*. Abingdon: Routledge, 2007.

Bunting, Madeleine. *The Plot: A Biography of an English Acre*. London: Granta, 2009.

Burnside, John. *The Good Neighbour*. London: Jonathan Cape, 2005.

Carson, Ciaran. *The Ballad of HMS Belfast: A compendium of Belfast poems*. Loughcrew: Gallery Press, 1999.

———. 'Belfast'. Box 27, Folder 5, MSS 746. *Ciaran Carson Papers*. Manuscript, Archives and Rare Books Library, Emory University, GA.

———. *Belfast Confetti*. Loughcrew: Gallery Press, 1989.

———. *Breaking News*. Loughcrew: Gallery Press, 2003.
———. 'Escaped from the massacre?'. *Honest Ulsterman* 50 (1975): 183–6.
———. *For All We Know*. Loughcrew: Gallery Press, 2008.
———. *The Irish for No*. Loughcrew: Gallery Press, 1987.
———. *Last Night's Fun: About Music, Food and Time*. London: Pimlico, 1997.
———. *The Star Factory*. London: Granta, 1997.
———. 'This is what libraries are for'. *The Dublin Review* 4 (2001): n.p. Web. 13 July 2011.
———. *West Belfast – The Falls, 1931 (Old Ordnance Survey Maps)*. Gateshead: Alan Godfrey, 1989.
Clark, Kenneth. *Landscape into Art*. London: Penguin, 1956.
Clark, Thomas A. *The Hundred Thousand Places*. Manchester: Carcanet, 2009.
Clark, Timothy. *The Cambridge Introduction to Literature and the Environment*. Cambridge: Cambridge University Press, 2011.
Clingman, Stephen. *The Grammar of Identity: Transnational Fiction and the Nature of the Boundary*. Oxford: Oxford University Press, 2009.
Collins, Lucy. '"Towards a Brink": The Poetry of Kathleen Jamie and Environmental Crisis'. *Crisis and Contemporary Poetry*. Ed. Anne Karhio, Seán Crossan, and Charles I. Armstrong. Basingstoke: Palgrave Macmillan, 2011. 150–66.
Cooke, Simon. *Travellers' Tales of Wonder: Chatwim, Naipaul, Sebald*. Edinburgh: Edinburgh University Press, 2013.
Corcoran, Neil. 'One Step Forward, Two Steps Back: Ciaran Carson's *The Irish for No*'. *The Chosen Ground: Essays on Contemporary Poetry of Northern Ireland*. Ed. Neil Corcoran. Bridgend: Seren, 1992. 211–33.
Cosgrove, Denis. 'Prospect, perspective and the evolution of the landscape idea'. *Transactions of the Institute of British Geographers* NS 10.1 (1985): 45–62.
Cosgrove, Denis, and Stephen Daniels, eds. *The Iconography of Landscape: Essays on the symbolic representation, design and use of past environments*. Cambridge: Cambridge University Press, 1988.
Coverley, Merlin. *Psychogeography*. Harpenden: Pocket Essentials, 2006.
Cowley, Jason. 'Editor's Letter'. *Granta 102: The New Nature Writing* (Summer 2008): 10–11.
Crang, Mike. *Cultural Geography*. London: Routledge, 1998.
Cresswell, Tim. 'Landscape and the Obliteration of Practice'. *Handbook of Cultural Geography*. Ed. Kay Anderson et al. London: Sage, 2003. 269–81.
———. *Place: a short introduction*. Oxford: Blackwell, 2004.
Cronon, William. 'Kennecot Journey'. *Under an Open Sky: Rethinking America's Western Past*. Ed. William Cronon, George Miles, and Jay Gitlin. New York: Norton, 1992. 28–51.
———. *Nature's Metropolis: Chicago and the Great West*. London: Norton, 1991.
———. 'A Place for Stories: Nature, History, and Narrative'. *The Journal of American History* 78.4 (1992): 1347–76.
Crouch, David. 'Flirting with Space: thinking landscape relationally'. *Cultural Geographies* 17 (2010): 5–18.

Cueppens, Jan. 'Tracing the Witness in W.G. Sebald'. *W.G. Sebald and the Writing of History*. Ed. Anne Fuchs and J.J. Long. Würzburg, Germany: Königshausen & Neumann, 2007. 59–72.
Curry, Patrick. 'Lost Ways'. *Times Literary Supplement* 8 February 2008: 26.
Daniels, Stephen. 'Marxism, culture and the duplicity of landscape'. *New Models in Geography*. 2 vols. Ed. N. Thrift and R. Peet. Boston: Allen and Unwin, 1990. 196–220.
Daniels, Stephen, et al., eds. *Envisioning Landscapes, Making Worlds: Geography and the Humanities*. Abingdon: Routledge, 2011.
Daniels, Stephen, and Hayden Lorimer. 'Until the end of days: narrating landscape and environment'. *Cultural Geographies* 19.1 (2012): 3–9.
Daniels, Stephen, Mike Pearson, and Heike Roms. 'Editorial'. *Performance Research* 15.4 (2010): 1–5.
Darby, David. 'Landscape and Memory: Sebald's Redemption of History'. *W.G. Sebald: History – Memory – Trauma*. Ed. Scott Denham and Mark McCulloh. Berlin: Walter de Gruyter, 2006. 265–77.
Davidson, Ian. *Ideas of Space in Contemporary Poetry*. Basingstoke: Palgrave, 2007.
———. *Radical Spaces of Poetry*. Basingstoke: Palgrave, 2010.
Deakin, Roger. *Wildwood: A Journey Through Trees*. London: Penguin, 2008.
Debord, Guy. 'Introduction to a Critique of Urban Geography'. *The Situationists and the City*. Ed. Tom McDonough. London: Verso, 2009. 59–63.
———. 'Theory of the *Dérive*'. *The Situationists and the City*. Ed. Tom McDonough. London: Verso, 2009. 77–85.
de Certeau, Michel. *The Practice of Everyday Life*. Trans. Stephen Rendall. Berkeley: University of California Press, 1984.
Denman, Peter. 'Language and the prosodic line in Carson's poetry'. *Ciaran Carson: Critical Essays*. Ed. Elmer Kennedy-Andrews. Dublin: Four Courts Press, 2009. 28–44.
Derrida, Jacques. *Demeure: Fiction and Testimony.* Trans. Elizabeth Rottenberg. Stanford, CA: Stanford University Press, 2000.
———. *Of Grammatology*. Trans. Gayatri Chakravorty Spivak. Baltimore, MD: Johns Hopkins University Press, 1997.
Dewsbury, John, et al. 'Enacting geographies'. *Geoforum* 33 (2002): 437–40.
Doob, Penelope Reed. *The Idea of the Labyrinth: from Classical Antiquity through the Middle Ages*. Ithaca: Cornell University Press, 1990.
Drabble, Margaret. *A Writer's Britain.* London: Thames & Hudson, 2009.
Dubow, Jessica. 'Out of Place, After Time: W.G. Sebald and Landscape as Loss'. *Literary Dynamics of Place*. University of Durham, 9 April 2011. Lecture.
———. 'Still-Life, After-Life, *Nature-Morte*; W.G. Sebald and the demands of landscape'. *Envisioning Landscapes, Making Worlds: Geography and the Humanities*. Ed. Stephen Daniels et al. Abingdon: Routledge, 2011. 188–97.
Ebbatson, Roger. *Landscape and Literature 1830–1914: Nature, Text, Aura.* Basingstoke: Palgrave, 2013.

Edensor, Tim. 'Walking Through Ruins'. *Ways of Walking: Ethnography and Practice on Foot*. Ed. Tim Ingold and Jo Lee Vergunst. Farnham: Ashgate, 2008. 123–41.

Evans, Gareth, and Di Robson, eds. *Towards Re-Enchantment: Place and Its Meanings*. London: Artevents, 2010.

Falci, Eric. 'Place, Space, and Landscape'. *A Concise Companion to Postwar British and Irish Poetry*. Ed. Nigel Alderman and C.D. Blanton. Oxford: Wiley-Blackwell, 2009. 200–220.

Faris, Wendy B. 'The Labyrinth as Sign'. *City Images: Perspectives from Literature, Philosophy and Film*. Ed. Mary Ann Caws. New York: Gordon and Breach, 1991. 33–41.

Farley, Paul, and Michael Symmons Roberts. *Edgelands: Journeys into England's True Wilderness*. London: Jonathan Cape, 2011.

Felman, Shoshana, and Dori Laub. *Testimony: Crises of Witnessing in Literature, Psychoanalysis, and History*. Abingdon: Routledge, 1992.

Ferguson, Christine. 'Victoria-Arcana and the Misogynistic Poetics of Resistance in Iain Sinclair's *White Chappell Scarlet Tracings* and Alan Moore's *From Hell*'. *LIT: Literature Interpretation Theory* 20 (2009): 45–64.

Fuchs, Anne, and J.J. Long, eds. *W.G. Sebald and the Writing of History*. Würzburg, Germany: Königshausen & Neumann, 2007.

Gairn, Louisa. *Ecology and Modern Scottish Literature*. Edinburgh: Edinburgh University Press, 2008.

Gee, Grant, dir. *Patience (After Sebald)*. London: Soda Films, 2012. DVD.

Gilbert, Roger. *Walks in the World: Representation and Experience in Modern American Poetry*. Princeton: Princeton University Press, 1991.

Gillis, Alan. 'Ciaran Carson: Beyond Belfast'. *The Cities of Belfast*. Ed. Nicholas Allen and Aaron Kelly. Dublin: Four Courts Press, 2003. 183–98.

Goodby, John. '"Walking in the City": space, narrative, and surveillance in *The Irish for No* and *Belfast Confetti*'. *Ciaran Carson: Critical Essays*. Ed. Elmer Kennedy-Andrews. Dublin: Four Courts Press, 2009. 66–85.

Goring, Rosemary. 'Kathleen Jamie: The SRB Interview'. *Scottish Review of Books* 8 June 2012: n.p. Web. 16 April 2014.

Groes, Sebastian. *The Making of London: London in Contemporary Literature*. Basingstoke: Palgrave Macmillan, 2011.

Harley, J.B. 'Deconstructing the Map'. *Writing Worlds: discourse, text and metaphor in the representation of landscape*. Ed. Trevor J. Barnes and James S. Duncan. London: Routledge, 1992. 231–47.

Harrison, Paul. 'The Caesura: remarks on Wittgenstein's interruption of theory, or, why practices elude explanation'. *Geoforum* 33 (2002): 487–503.

Hart, Matthew, and Tania Lown-Hecht. 'The Extraterritorial Poetics of W.G. Sebald'. *Modern Fiction Studies* 58.2 (2012): 214–38.

Harvey, David. *The Condition of Postmodernity: An Enquiry into the Origins of Cultural Change*. Oxford: Basil Blackwell, 1989.

Hazlitt, William. *Selected Essays*. Ed. George Sampson. Cambridge: Cambridge University Press, 1924.

Heidegger, Martin. *Basic Writings*. Ed. David Farrell Krell. London: Routledge, 1993.
———. *Being and Time*. Trans. John Macquerrie and Edward Robinson. Oxford: Blackwell, 1962.
Hones, Sheila. 'Text as It Happens: Literary Geography'. *Geography Compass* 2.5 (2008): 1301–17.
Houen, Alex. *Terrorism and Modern Literature*. Oxford: Oxford University Press, 2002.
Hunt, Stephen E. 'The Emergence of Psychoecology: The New Nature Writings of Roger Deakin, Mark Cocker, Robert Macfarlane and Richard Mabey'. *Green Letters* 10 (2009): 70–77.
Hutchinson, Ben. '"Egg boxes stacked in a crate": Narrative Status and its Implications'. *W.G. Sebald: History – Memory – Trauma*. Ed. Scott Denham and Mark McCulloh. Berlin: Walter de Gruyter, 2006. 171–82.
Ingold, Tim. *Lines: A Brief History*. London: Routledge, 2007.
———. *The perception of the environment: essays on livelihood, dwelling and skill*. Abingdon: Routledge, 2000.
Isherwood, Christopher. *Goodbye to Berlin*. London: Vintage, 1998.
Jackson, John Brinckerhoff. *Discovering the Vernacular Landscape*. New Haven: Yale University Press, 1984.
James, David. 'As He Stalks His "Pillar to the Dispossessed": Confessions of the Cryptographer'. *City Visions: The Work of Iain Sinclair*. Ed. Robert Bond and Jenny Bavidge. Newcastle: Cambridge Scholars Publishing, 2007. 148–61.
———. *Contemporary British Fiction and the Artistry of Space: Style, Landscape, Perception*. London: Continuum, 2008.
Jamie, Kathleen. 'Author Statement'. *British Council: Literature*. Web. 23 April 2014.
———. 'Diary'. *London Review of Books* 14 July 2011: 38–9.
———. *Findings*. London: Sort Of Books, 2005.
———. *Kathleen Jamie*. Web. 15 May 2014.
———. 'A Lone Enraptured Male'. *London Review of Books* 6 March 2008: 25–7.
———. 'On Rona'. *Towards Re-Enchantment: Place and Its Meanings*. Ed. Gareth Evans and Di Robson. London: Artevents, 2010. 135–51.
———. *The Overhaul*. London: Picador, 2012.
———. *Sightlines*. London: Sort Of Books, 2012.
———. *Sightlines: A Conversation with the Natural World*. New York: The Experiment, 2013.
———. *The Tree House*. London: Picador, 2004.
———. 'Truth and Change in Poetry'. *Strong Words: modern poets on modern poetry*. Ed. W.N. Herbert and Matthew Hollis. Tarset: Bloodaxe, 2000. 277–81.
Jarvis, Robin. *Romantic Writing and Pedestrian Travel*. Basingstoke: Palgrave, 1997.
Jenks, Chris. 'Watching your step: the history and practice of the flâneur'. *Visual Culture*. Ed. Chris Jenks. London: Routledge, 1995. 142–60.

Kennedy-Andrews, Elmer, ed. *Ciaran Carson: Critical Essays*. Dublin: Four Courts Press, 2009.

———. 'For all I know: Ciaran Carson in conversation with Elmer Kennedy-Andrews'. *Ciaran Carson: Critical Essays*. Ed. Elmer Kennedy-Andrews. Dublin: Four Courts Press, 2009. 13–27.

Kerrigan, John. 'The Country of the Mind'. *Times Literary Supplement* 11 September 1998: 3–4.

Laing, Olivia. *To the River: A Journey Beneath the Surface*. Edinburgh: Canongate, 2011.

Laing, R.D. *The Divided Self: An Existential Study in Sanity and Madness*. London: Routledge, 1999.

Lamb, Charles. *Selected Prose*. Ed. Adam Philips. London: Penguin, 1985.

Lando, Fabio. 'Fact and fiction: Geography and literature; A bibliographic survey'. *Geojournal* 38.1 (1996): 3–18.

Landry, Donna. *The Invention of the Countryside: Hunting, Walking and Ecology in English Literature, 1671–1831*. Basingstoke: Palgrave, 2001.

Landy, Joshua, and Michael Saler, eds. *The Re-Enchantment of the World*. Stanford, CA: Stanford University Press, 2009.

Lefebvre, Henri. *The Production of Space*. Trans. Donald Nicholson-Smith. Oxford: Blackwell, 1991.

Long, J.J. 'W.G. Sebald's Miniature Histories'. *W.G. Sebald and the Writing of History*. Ed. Anne Fuchs and J.J. Long. Würzburg, Germany: Königshausen & Neumann, 2007. 111–20.

Long, J.J., and Anne Whitehead, eds. *W.G. Sebald: a critical companion*. Edinburgh: Edinburgh University Press, 2004.

Lorimer, Hayden. 'Cultural Geography: the busyness of being "more-than-representational"'. *Progress in Human Geography* 29 (2005): 83–94.

Lorimer, Hayden, and John Wylie. 'LOOP (a geography)'. *Performance Research* 15.4 (2010): 6–13.

Luckhurst, Roger. 'The contemporary London Gothic and the limits of the "spectral turn"'. *Textual Practice* 16 (2002): 527–46.

Mabey, Richard. *Beechcombings: The Narratives of Trees*. London: Vintage, 2008.

———. *A Good Parcel of English Soil*. London: Penguin, 2013.

———. 'On the Virtues of Dis-Enchantment'. *Towards Re-Enchantment: Place and Its Meanings*. Ed. Gareth Evans and Di Robson. London: Artevents, 2010. 29–38.

———. *Turned Out Nice Again: On Living With the Weather*. London: Profile Books, 2013.

———. *The Unofficial Countryside*. London: Sphere Books, 1978.

Macfarlane, Robert. 'Common Ground'. *Guardian* 4 June 2005: n.p. Web. 18 June 2010.

———. 'A Counter-Desecration Phrasebook'. *Towards Re-Enchantment: Place and Its Meanings*. Ed. Gareth Evans and Di Robson. London: Artevents, 2010. 107–30.

———. 'Foreword'. *A Wilder Vein*. Ed. Linda Cracknell. Ullapool: Two Ravens Press, 2009. vii–ix.
———. 'Introduction'. *Nature Near London*. By Richard Jefferies. London: Collins, 2012. xi–xxv.
———. 'Meshed and weathered'. *Guardian* 16 April 2011: Review 14.
———. *Mountains of the Mind: A History of a Fascination*. London: Granta, 2003.
———. *The Old Ways: A Journey on Foot*. London: Hamish Hamilton, 2012.
———. 'Very ordinary, very beautiful'. *Guardian* 25 September 2010: Review 16–17.
———. 'Way-rights'. *Archipelago* 6 (Winter 2011): 8–22.
———. *The Wild Places*. London: Granta, 2007.
Macfarlane, Robert, Stanley Donwood, and Dan Richards. *Holloway*. London: Faber, 2012.
Mallory, William E., and Paul Simpson-Housley, eds. *Geography and Literature: A Meeting of the Disciplines*. Syracuse, NY: Syracuse University Press, 1987.
Malpas, J.E. *Place and Experience: A Philosophical Topography*. Cambridge: Cambridge University Press, 1999.
Marks, John. 'W.G. Sebald: Invisible and Intangible Forces'. *New Formations* 55 (2005): 89–103.
Massey, Doreen. *For Space*. London: Sage, 2005.
———. 'Landscape/space/politics: an essay'. *The Future of Landscape*. Web. 27 July 2011.
Matless, David. 'Nature Voices'. *Journal of Historical Geography* 35 (2009): 178–88.
———. 'The Properties of Landscape'. *Handbook of Cultural Geography*. Ed. Kay Anderson et al. London: Sage, 2003. 227–32.
McCarthy, Conor. 'Ciaran Carson's Labyrinths'. *English* 54 (2005): 99–116.
McCracken, Kathleen. 'Ciaran Carson: Unravelling the Conditional, Mapping the Provisional'. *Poetry in Contemporary Irish Literature*. Ed. Michael Kinneally. Gerrards Cross: Colin Smythe, 1995. 356–72.
McCulloh, Mark. 'Introduction; Two Languages, Two Audiences: The Tandem Literary Oeuvres of W.G. Sebald'. *W.G. Sebald: History – Memory – Trauma*. Ed. Scott Denham and Mark McCulloh. Berlin: Walter de Gruyter, 2006. 7–20.
Merleau-Ponty, Maurice. *Phenomenology of Perception*. Trans. Colin Smith. Abingdon: Routledge, 2002.
Moritz, Charles P. *Travels, Chiefly on Foot, through Several Parts of England, in 1782*. London: G.G. & J. Robinson, 1795.
Moshenka, Joe. 'The energy of the walker, the absorption of the passenger'. *Literary London: Interdisciplinary Studies in the Representation of London* 3.2 (2005): n.p. Web. 14 July 2010.
Murray, Alex. *Recalling London: Literature and History in the work of Peter Ackroyd and Iain Sinclair*. London: Continuum, 2007.
Newman, Karen. *Cultural Capitals: Early Modern London and Paris*. Princeton: Princeton University Press, 2007.

Nicholson, Geoff. *The Lost Art of Walking: The History, Science, Philosophy, and Literature of Pedestrianism*. New York: Riverhead, 2008.

Nora, Pierre. 'Between Memory and History: *Les Lieux de Mémoire*'. *Representations* 26 (1989): 7–24.

Obert, Julia C. 'Sounding the city: Ciaran Carson and the perceptual politics of war'. *Textual Practice* 26 (2012): 1081–110.

Ogborn, Miles. 'Mapping Words'. *New Formations* 57 (2005/6): 145–9.

Ormsby, Frank. 'Ciaran Carson interviewed by Frank Ormsby'. *Linen Hall Review* (April 1991): 4–8.

Pearson, Mike. *In Comes I: Performance, Memory, and Landscape*. Exeter: University of Exeter Press, 2006.

Perril, Simon. 'A cartography of absence: the work of Iain Sinclair'. *Comparative Criticism* 19 (1997): 309–39.

Phillips, Richard, and Scott McCracken. 'Editorial'. *New Formations* 57 (2005/6): 7–9.

Pile, Steve. 'Memory and the City'. *Temporalities, autobiography and everyday life*. Ed. Jan Campbell and Janet Harbord. Manchester: Manchester University Press, 2002. 111–27.

Pinder, David. 'Arts of urban exploration'. *Cultural Geographies* 12 (2005): 383–411.

Pocock, Douglas C.D. 'Geography and literature'. *Progress in Human Geography* 12 (1988): 87–102.

———, ed. *Humanistic Geography and Literature: Essays on the Experience of Place*. London: Croom Helm, 1981.

Price, Patricia L. 'Cultural geography and the stories we tell ourselves'. *Cultural Geographies* 17.2 (2010): 203–10.

Richardson, Douglas. 'Converging worlds: geography and the humanities'. *Envisioning Landscapes, Making Worlds: Geography and the Humanities*. Ed. Stephen Daniels et al. Abingdon: Routledge, 2011. xix–xxi.

Richardson, Douglas, et al., eds. *Geohumanities: Art, history, text at the edge of place*. Abingdon: Routledge, 2011.

Robinson, Jeffrey C. *The Walk: Notes on a Romantic Image*. Norman: University of Oklahoma Press, 1989.

Ross, Kristin. *The Emergence of Social Space*. London: Verso, 2008.

Sanders, Julie. 'Beyond the power of art to represent? Narratives and performances of the Arctic in the 1630s'. *Envisioning Landscapes, Making Worlds: Geography and the Humanities*. Ed. Stephen Daniels et al. Abingdon: Routledge, 2011. 293–8.

Saunders, Angharad. 'Literary Geography: reforging the connections'. *Progress in Human Geography* 34.4 (2010): 436–52.

Schama, Simon. *Landscape and Memory*. London: HarperCollins, 1995.

Scott, Kirsty. 'In the nature of things'. *Guardian* 18 June 2005: n.p. Web. 27 January 2011.

Sears, John. '"Ghosts and Texts and Photographs": Sinclair's Overwriting of the Dead'. *City Visions: The Work of Iain Sinclair*. Ed. Robert Bond and Jenny Bavidge. Newcastle: Cambridge Scholars Publishing, 2007. 162–74.
Sebald, W.G. *After Nature*. Trans. Michael Hamburger. London: Penguin, 2002.
———. *Austerlitz*. Trans. Anthea Bell. London: Penguin, 2002.
———. *Campo Santo*. Trans. Anthea Bell. London: Penguin, 2006.
———. *The Emigrants*. Trans. Michael Hulse. London: Vintage, 2002.
———. *A Place in the Country*. Trans. Jo Catling. London: Hamish Hamilton, 2013.
———. *Die Ringe des Saturn: eine englische Wallfahrt*. Frankfurt am Main, Germany: Eichborn, 1995.
———. *The Rings of Saturn*. Trans. Michael Hulse. London: Vintage, 2002.
———. *Vertigo*. Trans. Michael Hulse. London: Vintage, 2002.
Self, Will. 'Incidents along the road'. *Guardian* 7 February 2009: Review 2–4.
Sheppard, Robert. *Iain Sinclair*. Tavistock: Northcote, 2007.
Shields, David. *Reality Hunger: A Manifesto*. London: Hamish Hamilton, 2010.
Shklovsky, Victor. 'Art as Technique'. *Russian Formalist Criticism: Four Essays*. Ed. Lee T. Lemon and Marion J. Reis. Lincoln: University of Nebraska Press. 3–24.
Sinclair, Iain. 'At the Edge of the Edge: Blue Fence, Yellow Mud'. *Literary London 2008: Representations of London in Literature*. Brunel University, 2 July 2008. Lecture.
———. *Edge of the Orison: In the Traces of John Clare's 'Journey Out of Essex'*. London: Penguin, 2006.
———. *Ghost Milk: Calling Time on the Grand Project*. London: Hamish Hamilton, 2011.
———. *Hackney, That Rose-Red Empire*. London: Penguin, 2010.
———. *Lights Out for the Territory: 9 Excursions in the Secret History of London*. London: Penguin, 2003.
———. *London Orbital: A Walk around the M25*. London: Penguin, 2003.
———. 'Water Walks'. *Towards Re-Enchantment: Place and Its Meanings*. Ed. Gareth Evans and Di Robson. London: Artevents, 2010. 13–27.
Smith, Jos. 'An archipelagic literature: re-framing "The New Nature Writing"'. *Green Letters: Studies in Ecocriticism* 17.1 (2013): 5–15.
Smith, Peter. *The Syntax of the City*. London: Hutchinson, 1977.
Snyder, Gary. *The Practice of the Wild*. Berkeley, CA: Counterpoint, 1990.
Soja, Edward. *Postmodern Geographies: The Reassertion of Space in Critical Social Theory*. London: Verso, 1989.
Solnit, Rebecca. *Wanderlust: A History of Walking*. London: Verso, 2002.
Sprackland, Jean. *Strands: A Year of Discoveries on the Beach*. London: Jonathan Cape, 2012.
Stainer, Jonathan. 'The possibility of non-sectarian futures: emerging disruptive identities of place in the Belfast of Ciaran Carson's *The Star Factory*'. *Environment and Planning D: Society and Space* 23 (2005): 373–94.

Stenning, Anna. 'An interview with Robert Macfarlane'. *Green Letters: Studies in Ecocriticism* 17 (2013): 77–83.

Stenning, Anna, and Terry Gifford. 'Editorial: Twentieth-century nature writing in Britain and Ireland'. *Green Letters: Studies in Ecocriticism* 17 (2013): 1–4.

Stephen, Leslie. *Studies of a Biographer*. 3 vols. London: Duckworth, 1902.

Stopford, J., ed. *Pilgrimage Explored*. Woodbridge: York Medieval Press, 1999.

Tally Jr., Robert T. *Spatiality*. Abingdon: Routledge, 2013.

Tarlo, Harriet, ed. *The Ground Aslant: An Anthology of Radical Landscape Poetry*. Exeter: Shearsman Books, 2011.

Terentowicz-Fotyga, Urszula. 'Unreal City to City of Referents: Urban Space in Contemporary London Novels'. *Journal of Narrative Theory* 39 (2009): 305–29.

Thacker, Andrew. 'The Idea of a Critical Literary Geography'. *New Formations* 57 (2005/6): 56–73.

———. 'Journey with Maps: Travel Theory, Geography and the Syntax of Space'. *Cultural Encounters: European Travel Writing in the 1930s*. Ed. Charles Burdett and Derek Duncan. Oxford: Berghahn Books, 2002. 11–28.

———. *Moving Through Modernity: Space and Geography in Modernism*. Manchester: Manchester University Press, 2009.

Theisen, Bianca. 'Prose of the World: W.G. Sebald's Literary Travels'. *Germanic Review* 79 (2004): 163–79.

Thrift, Nigel. *Spatial Formations*. London: Sage, 1996.

Tuan, Yi-Fu. *Space and Place: The Perspective of Experience*. London: Edward Arnold, 1977.

Turner, Victor. *From Ritual to Theatre: The Human Seriousness of Play*. New York: PAJ Publications, 1982.

———. *The Ritual Process: Structure and Anti-Structure*. New York: Aldine de Gruyter, 1995.

Wallace, Anne D. *Walking, Literature, and English Culture: The Origins and Uses of Peripatetic in the Nineteenth Century*. Oxford: Clarendon, 1994.

Walsham, Alexandra. *The Reformation of the Landscape: Religion, Identity, and Memory in Early Modern Britain and Ireland*. Oxford: Oxford University Press, 2011.

Weber, Max. *From Max Weber: Essays in Sociology*. Ed. Hans Gerth and C. Wright Mills. Oxford: Oxford University Press, 1946.

Westphal, Bertrand. *Geocriticism: Real and Fictional Spaces*. Trans. Robert T. Tally Jr. New York: Palgrave Macmillan, 2011.

Wheatley, David. '"Pushed next to nothing": Ciaran Carson's *Breaking News*'. *Ciaran Carson: Critical Essays*. Ed. Elmer Kennedy-Andrews. Dublin: Four Courts Press, 2009. 45–65.

———. '"That blank mouth": Secrecy, shibboleths, and silence in Northern Irish poetry'. *Journal of Modern Literature* 25.1 (2001): 1–16.

Williams, Raymond. *The Country and the City*. Oxford: Oxford University Press, 1975.

———. *Keywords: A vocabulary of culture and society*. London: Fontana, 1988.
Winspur, Steven. 'On City Streets and Narrative Logic'. *City Images: Perspectives from Literature, Philosophy and Film*. Ed. Mary Ann Caws. New York: Gordon & Breach, 1991. 60–70.
Wittgenstein, Ludwig. *Philosophical Investigations*. Trans. G.E.M. Anscombe. Oxford: Blackwell, 1997.
Wolfreys, Julian. 'Londonography: Iain Sinclair's Urban Graphic'. *Literary London: Interdisciplinary Studies in the Representation of London* 3.2 (2005): n.p. Web. 14 July 10.
———. *Writing London: The Trace of the Urban Text from Blake to Dickens*. Basingstoke: Palgrave, 1998.
———. *Writing London, Volume 2: Materiality, Memory, Spectrality*. Basingstoke: Palgrave Macmillan, 2004.
———. *Writing London, Volume 3: Inventions of the City*. Basingstoke: Palgrave Macmillan, 2007.
Worpole, Ken. 'East of Eden'. *Towards Re-Enchantment: Place and Its Meanings*. Ed. Gareth Evans and Di Robson. London: Artevents, 2010. 61–81.
Worpole, Ken (text), and Jason Orton (images). *The New English Landscape*. London: Field Station, 2013.
Wylie, John. 'Becoming-icy: Scott and Amundsen's South Polar voyages, 1910–1913'. *Cultural Geographies* 9 (2002): 249–65.
———. *Landscape*. Abingdon: Routledge, 2007.
———. 'A single day's walking: narrating self and landscape on the South West Coast Path'. *Transactions of the Institute of British Geographers* NS 30 (2005): 234–47.
———. 'The spectral landscapes of W.G. Sebald'. *Cultural Geographies* 14 (2007): 171–88.
———. 'Writing through Landscape'. *Process: Landscape and Text*. Ed. Catherine Brace and Adeline Johns-Putra. Amsterdam, Netherlands: Rodopi, 2010. 45–64.

# Index

For Product Safety Concerns and Information please contact our EU
representative GPSR@taylorandfrancis.com
Taylor & Francis Verlag GmbH, Kaufingerstraße 24, 80331 München, Germany

www.ingramcontent.com/pod-product-compliance
Ingram Content Group UK Ltd.
Pitfield, Milton Keynes, MK11 3LW, UK
UKHW022123190926
475702UK00013B/1617

* 9 7 8 0 3 6 7 1 4 0 5 3 3 *